T0270998

CLEAN WATER POLICY AND STATE CHOICE

The Water Quality Act of 1987 ushered in a new era of clean water policy to the Un. States. The Act stands today as the longest-lived example of national water quality poli It included a then-revolutionary funding model for wastewater infrastructure – the Cle Water State Revolving Fund – which gave states much greater authority to allocate clea water infrastructure resources. Significant differences between states exist in terms of their ability to provide adequate resources for the program, as well as their ability (or willingness) to meet the wishes of Congress to serve environmental needs and communities. This book examines the patterns of state program resource distribution using case studies and analysis of state and national program data. This book is important for researchers from a range of disciplines, including water, environmental and infrastructure policy, federalism/intergovernmental relations, intergovernmental administration, and natural resource management, as well as policy makers and policy advocates.

John C. Morris is a professor in the Department of Political Science at Auburn University. He has studied environmental policy and water policy for more than twenty-five years, and has published widely in public administration and public policy. He is the co-editor of *Speaking Green with a Southern Accent: Environmental Management and Innovation in the South* (2010), and *True Green: Executive Effectiveness in the US Environmental Protection Agency* (2012). He is co-editor of *Building the Local Economy: Cases in Economic Development* (2008); co-editor of a three-volume series *Prison Privatization: The Many Facets of a Controversial Industry* (2012); and *Advancing Collaboration Theory: Models, Typologies, and Evidence* (2016). His most recent books include *The Case for Grassroots Collaboration: Social Capital and Ecosystem Restoration at the Local Level* (2013, with others); *State Politics and the Affordable Care Act: Choices and Decisions* (Routledge, 2019, with others); and *Organizational Motivation for Collaboration: Theory and Evidence* (2019, with Diaz-Kope). In addition, he has published more than 130 peer-reviewed articles, book chapters, and reports.

CLEAN WATER POLICY AND STATE CHOICE

Promise and Performance in the Water Quality Act

JOHN CHARLES MORRIS

Auburn University

FOREWORD BY A. STANLEY MEIBURG

CAMBRIDGE
UNIVERSITY PRESS

CAMBRIDGE
UNIVERSITY PRESS

University Printing House, Cambridge CB2 8BS, United Kingdom

One Liberty Plaza, 20th Floor, New York, NY 10006, USA

477 Williamstown Road, Port Melbourne, VIC 3207, Australia

314–321, 3rd Floor, Plot 3, Splendor Forum, Jasola District Centre, New Delhi – 110025, India

103 Penang Road, #05–06/07, Visioncrest Commercial, Singapore 238467

Cambridge University Press is part of the University of Cambridge.

It furthers the University's mission by disseminating knowledge in the pursuit of education, learning, and research at the highest international levels of excellence.

www.cambridge.org
Information on this title: www.cambridge.org/9781108839129
DOI: 10.1017/9781108989138

First published 2022

A catalogue record for this publication is available from the British Library.

Library of Congress Cataloging-in-Publication Data
Names: Morris, John C. (John Charles), 1959- author. | Meiburg, A. Stanley, author of foreword.
Title: Clean water policy and state choice : promise and performance in the Water Quality Act /
John C. Morris ; foreword by A. Stanley Meiburg.
Other titles: Promise and performance in the Water Quality Act
Description: [Cambridge] : [Cambridge University Press], [2022] | Includes bibliographical references and index.
Identifiers: LCCN 2021031754 (print) | LCCN 2021031755 (ebook) | ISBN 9781108839129 (hardback) |
ISBN 9781108989138 (epub)
Subjects: LCSH: United States. Water Quality Act of 1987. | Water–Pollution–Laws and legislation–United States.
| Water quality management–United States.
Classification: LCC K3585.6 M67 2021 (print) | LCC K3585.6 (ebook) | DDC 344.7304/6343–dc23
LC record available at https://lccn.loc.gov/2021031754
LC ebook record available at https://lccn.loc.gov/2021031755

ISBN 978-1-108-83912-9 Hardback

To Betsy

The sewer is the conscience of the city. All things converge into it and are confronted with one another.

Victor Hugo

Contents

Figures

Tables

Foreword

Seen from the perspective of the fractured politics of 2021, the building of America's modern environmental statutory architecture seems almost miraculous. The period between the 1969 National Environmental Policy Act and the 1990 Clean Air Act Amendments created the bedrock of environmental protection in America. No less than twenty major pieces of federal legislation became law, covering eleven major areas and giving the new Environmental Protection Agency (EPA) an overwhelming agenda. It is tempting to look back at this development as inevitable, when in fact each piece of legislation faced a myriad of competing interests and scientific, legal, and financial controversies.

The sweeping authorities and ambitions of the 1972 Clean Water Act, the drama of its enactment over President Nixon's veto, and the nation's inability to meet its bold promises led to efforts during the following two decades to reconcile its aspirations with available technologies, resources, and competing national priorities. The 1987 Water Quality Act, the subject of this book, came between the dramatic reauthorization of Superfund the previous year and the all-consuming quest for Clean Air Act Amendments in 1990, and is sometimes overlooked. But this neglect is undeserved, and Professor Morris here gives us a timely correction.

While the 1987 Water Quality Act is built on its 1972 foundation, it embodies significant differences in approach. Professor Morris notes that underlying these differences are competing framings of federalism, the central tension of which is the desire for both national consistency and local flexibility. The 1972 Clean Water Act leaned toward federally directed construction grant expenditures in support of a distinct national goal. The 1987 Water Quality Act shifted this framing to give states more authority and discretion in determining the ultimate beneficiaries of government financial support for water quality objectives. Following the classic Harold Lasswell definition of politics as who gets what, when, and how, this shift is a profound one.

While the policy embedded in the 1987 Act may fairly be said to embody "Reagan federalism," the namesake of the term evidently did not think so, as the history of 1972 repeated itself when President Reagan vetoed the 1987 Act only to be overridden by Congress. It is noteworthy that no major environmental legislation since then has been adopted over a presidential veto.

It is also noteworthy that the 1987 Act reflected recommendations from a 1984 EPA task force, initiated by EPA Administrator William Ruckelshaus. This collaboration between the Executive Branch and the Congress would recur in the development of the 1990 Clean Air Act Amendments, but has become less common since then as partisan tensions have increased. Still, it illustrates a larger point: Effective legislative change is more likely when there is close collaboration between the two branches, or at least between committee and agency staff. Even now, it still works: The adoption of the thirty-year loan repayment authorization in the 2014 Water Resources Development Act followed recommendations from the EPA.

The 1987 Act has remained essentially the same as when it first became law, and largely did what its authors intended it to do. Not only did the 1987 Act shift the responsibility for paying for municipal water infrastructure to mechanisms which gave states both more financial control and more financial responsibility, it also shifted the burden of paying for these improvements to future generations who would benefit from them (via loan repayments). In an era where federal budget deficits were a prominent political concern, it created a mechanism to finance continuing improvements through a revolving fund that in theory, once capitalized by the federal government, would be self-sustaining and not in constant need of federal appropriations.

What more can we say about how the 1987 Water Quality Act has worked? Like most ambitious legislation, there are multiple answers to this deceptively simple question. Professor Morris expertly examines how these answers depend on different understandings of what "worked" means. From a financial standpoint, by leveraging the impact of federal appropriations, the 1987 Act has succeeded in making more resources available to address pollution discharges from municipal sewers and treatment works, though federal capitalization grants to the State Revolving Loan Funds have continued far beyond the time that the 1987 Act envisioned or authorized. From an environmental standpoint, most observers agree that water quality in the United States has improved since the early 1980's, though precise measures of this at a national level are hard to come by.

A closer look reveals subtler underlying tensions. One revolves around the enduring "who pays" question. Professor Morris notes that many state and local governments used the construction grants program established in the 1972 Clean Water Act as a substitute for, rather than as a supplement to, their own investments in wastewater infrastructure. The construction grants program was supposed to address existing pollution in 1972; communities were supposed to include the cost of treatment for growth in their planning, with new treatment facilities subject to rigorous technology standards. Indeed, the belief that construction grants were being used to promote economic development was an underlying driver behind the effort in the 1987 amendments to shift more of the financial responsibility for clean water back to state and local governments.

Yet the desire of the authors of the 1987 Act to reduce claims on the federal budget created no political incentive for Congress or the states to abandon continuing federal funding for water and sewer infrastructure. As Professor Morris observes, EPA surveys for both drinking water and wastewater treatment have continued to display national needs well in excess of the amounts available in the revolving loan funds. Capitalization grants have continued to the present day, with bipartisan support, far beyond the original end date of fiscal year 1994. As of this writing, roughly $2 billion per year, a quarter of EPA's total budget, goes to such grants.

The revolving loan concept expanded in 1996 to drinking water through amendments to the Safe Drinking Water Act, so that drinking water supply, historically even more of a local responsibility than wastewater treatment, is now also a beneficiary of federal support. The principle of state flexibility has been expanded by allowing states to shift funds between the two revolving loan accounts, and use loan forgiveness provisions more widely. In 2009 the funds became instruments of economic recovery, receiving an additional $6 billion in the American Recovery and Reinvestment Act. More federal support came in 2014 through the Water Infrastructure Finance and Innovation Act, which made funds available for federal loan guarantees for creditworthy projects. As of July 2020, EPA had provided about $5.3 billion in loan assistance through this mechanism, with an additional $6 billion anticipated in the coming year (EPA 2021a).

This is not to say that the revolving loan concept has failed; indeed, it has been successful in making more revenue available than grants alone would have done. For the Clean Water State Revolving Loan Fund (CWSRF), EPA estimates that through 2019, federal capitalization grants of $45.2 billion have resulted in 41,234 low-interest loans totaling $138 billion (EPA 2021b). Projects available for support by such loans have expanded to include such things as energy efficiency enhancements at treatment plants. Net assets of the Clean Water SRF at the end of FY 2019 exceeded $53 billion (EPA 2019). States have become more adept as financial managers, and the conservative engineers who managed SRF accounts in many states have been supplanted by more sophisticated financial managers, enabling greater leveraging of SRF assets.

However, critical issues remain, especially those of affordability, emerging contaminants, and governance. Affordability is a constant concern, not just for capital expenses but for operation and maintenance. Just as individuals can find themselves "house poor," capital-intensive projects, encouraged by subsidized financing, can outstrip the financial ability of communities if poorly thought through. Deferred maintenance, leading to such problems as sanitary sewer overflows and infiltration/inflow, remains a problem for many systems. Communities that are small or have a large number of lower-income households may not be able to meet matching share requirements or manage even subsidized loans. "Affordability" is more of a household issue than a community issue, but wastewater utilities tend to resist any fee structure other than a pay-per-use model, and full cost pricing also meets resistance from local taxpayers. The result of all this is that there are still areas in America where basic sewage treatment is not available.

Emerging contaminants pose a special challenge. Evidence has grown in recent years that the conventional elements of wastewater pollution – pH, temperature, suspended solids, bacteria, and nutrients – are being joined by such things as low concentrations of pharmaceuticals, per- and polyfluoroalkyl substances (PFAS), and organic chemicals such as 1, 4-dioxane. Technologies to remove such things at treatment plants are expensive, and are best addressed by source reduction.

Governance remains a problem in many places. In the drinking water area (where records are more complete), the American Society of Civil Engineers reported that in 2017 there were about 155,000 active public drinking water systems across the country (ASCE 2017). Community water systems comprised about 51,000 of this total, and 17 percent of these serve about 92 percent of the total US population. The remaining small community systems serve about 8 percent of the population, and frequently lack both economies of scale and financial, managerial, and technical capacity. The same dynamics are at work in wastewater systems. Consolidations can help address these shortcomings, but they challenge community identities and face political obstacles.

Competition between the SRFs and the private capital market has also appeared in ways that seemed unlikely when Congress passed the 1987 Act. High interest rates in the mid-1980's made the need for loan subsidies more compelling, and borrowers were more willing to accept the special conditions that came with accepting federal funds (e.g., Buy American requirements, Davis-Bacon wage rates). However, as interest rates declined during the Great Recession and stayed low through the ensuing recovery, creditworthy systems found they could increasingly raise capital for improvements in the private bond market at terms comparable to and/or with fewer transaction costs than SRF loans. The federal government contributed to this development through one tool that the 1987 Water Quality Act did not change: vigorous use by EPA of its enforcement authority. Thanks to a national targeting strategy, scores of cities around the United States found themselves the subject of federal lawsuits between 1990 and 2015.

The settlement agreements resulting from these lawsuits were not conditioned on the federal government paying for injunctive relief.

Finally, since 1987, water quality objectives have become stronger drivers of the national program. Lawsuits over the Clean Water Act's requirements for total maximum daily loads have focused attention on the effects of nonpoint sources in creating water quality impairments. But these nonpoint sources are hard to quantify, and point sources, both municipal and industrial, have felt themselves to be under disproportionate scrutiny because their effluents could be measured and monitored, whereas others could not. Professor Morris finds that there has been some movement to use SRF resources to address nonpoint source, but much work still remains.

All of this reflects how federalism means different things to different actors. To some it is a bedrock philosophy of government, to some a legal/constitutional issue, to some a pragmatic division of authority and responsibility, and to some simply a means to an end – a matter of expedience in achieving policy objectives. This ambiguity goes back to the framers of the Constitution, and we tolerate it out of a belief that in a country as big and diverse as the United States, the benefits of state discretion outweigh the costs of inconsistent administration. But this choice means that unequal water quality outcomes for citizens are virtually guaranteed. Some states make more effective use of federal resources than others; some states use resources more wisely than the federal government would in their place; some states have policy priorities that differ from ones that Congress has set; some states lag behind.

The nation continues to grapple with the financial, environmental, and governance challenges raised by the 1987 Water Quality Act. As this book goes to press, debates about these challenges are in full bloom. The massive infrastructure bill passed by the Senate in August 2021 contains provisions that would amend the Clean Water Act to add far more money to the system and create more opportunities for loan forgiveness – essentially grants – to disadvantaged communities. In addition, although it is still too early to be certain, it appears that the bill may give EPA a stronger role relative to states in directing such investments. We have seen genuine progress in protecting American water quality, yet we have fallen short of the ambitious 1972 goals of fishable and swimmable waters everywhere. Professor Morris has done us a great service in telling the story of how our present condition came to pass, what has worked as expected and what has not, and how difficult the path ahead remains.

A. Stanley Meiburg

Director, Graduate Studies in Sustainability, Wake Forest University, and former Acting Deputy Administrator, United States Environment Protection Agency

Acknowledgments

This book was more than thirty years in the making. I began study of the Water Quality Act as a new doctoral student at Auburn University in 1989. Indeed, the reason I entered the Ph.D. program at the time was the promise of a graduate assistantship to work on a large grant funded by the US Geological Survey (USGS) to study the early implementation of the Clean Water State Revolving Fund (CWSRF) program. The grant recipients, Dr. John G. Heilman and Dr. Gerald W. Johnson, brought me on board their project more in the role of a full team member than as a Graduate Research Assistant. Along with Dr. Laurence J. O'Toole Jr., these three provided an unparalleled research experience, and instilled in me both a passion and curiosity that I strive to pay forward to my students each day. I am indebted to them for their guidance, mentorship, and friendship.

There were others that were part of the team as well; they were instrumental in the early work. My fellow Ph.D. Graduate Research Assistant and doctoral student, Dr. Gloria (Burch) Smith, was an engaged and thoughtful colleague, and together we wrote several conference papers that formed the foundation for much of my later work with this program. In addition, we had several MPA students that worked with us at various times, including Rod Bramblett, DeRon Decker, and Robert W. Gottesman, each of whom provided assistance coding much of the early program data. Professors Johnson and Heilman, along with Robert Bernstein and Thomas Vocino, served as my dissertation committee; there is a direct line between that dissertation and this book.

In the intervening years, a number of coauthors on various conference papers and journal articles also contributed to the knowledge base. These include Rick Travis (Mississippi State University), Elizabeth D. Morris (US Government Accountability Office), Gerald A. Emison (Mississippi State University), Ryan Williamson and Jonathan Fisk (Auburn University), Martin Mayer (University of North Carolina–Pembroke), Jan C. Hume (Auburn University), Lien Nguyen (Old Dominion University), and David Breaux (Delta State University). Their contributions, large and small, are an integral part of this book, and those contributions are gratefully acknowledged. I am also indebted to a number of very able graduate assistants over the years, including Byron Price, Tom Poulin, Jan Hume, Lien Nguyen, Brian Ezenou, Robert Summers, Benjamin McGarr, Daryl Jones, and Jennifer (Jordan) Blevins.

In addition, there were a number of people who provided invaluable assistance and guidance through the preparation of this manuscript. Ryan Williamson offered outstanding tutelage and guidance for the statistical analyses for this study, and his expertise, patience, and unfailing willingness to guide me through the intricacies of both Stata programming and statistical interpretation are greatly appreciated. Any errors of analysis or interpretation are mine, not his. Ryan also produced the maps presented in this book, and reviewed the empirical chapters in this volume. Jerry Emison, a longtime friend and colleague, volunteered for the enormous task of reading the entire draft manuscript and providing comments, and his efforts are greatly appreciated. His insights helped strengthen the manuscript considerably. Jonathan Fisk provided valuable advice regarding the structure of the manuscript, and was always willing to serve as a "sounding board" for different ideas and approaches. Shana Campbell Jones reviewed the discussion on the legal aspects of the Clean Water Act and the Water Quality Act. Soren Jordan generously provided technical assistance with Qualtrics. A special thanks is due to Lien Nguyen, who provided invaluable assistance to translate the EPA data from pdf files to spreadsheets for analysis. During the winter of 2018, I was on sabbatical from Old Dominion University; Cynthia Bowling, then Chair of the Department of Political Science at Auburn, kindly invited me to spend time working at Auburn, a period during which much of the early development work on this manuscript was completed. I am grateful for her encouragement and support.

In addition, several anonymous reviewers enlisted by Cambridge University Press provided outstanding reviews, suggestions, and comments that collectively strengthened the manuscript in very substantial ways, and I am indebted to them for their time and expertise. My editor at Cambridge University Press, Matthew Lloyd, was enthusiastic about the project from the outset, and he and his outstanding team at Cambridge University Press, including E. A. Sarah Lambert, Content Manager Sapphire Deveau, Project Manager Siddharthan Indra Priyadarshini, and Litty Santosh, the copy editor assigned to this manuscript, helped to ensure that the publication process was both smooth and trouble-free.

I also wish to thank James Giattina and Alexis Strauss, USEPA (retired), for their insights and observations. Both Jim and Alexis were very gracious with both their time and expertise, and both are much appreciated. A. Stanley Meiburg, former acting deputy administrator of USEPA, was both an early supporter of this project, and was gracious with both his time and his willingness to make introductions and contacts at EPA. The observant reader will also note that Dr. Meiburg also wrote the Foreword for this book, for which I am also very grateful.

I am deeply indebted to my late parents, Charles H. and Mary B. Morris, for their love and support; for instilling in me a curiosity about the world; and for providing me an increasingly rare opportunity: a debt-free education. Finally, I am indebted to my wife, Betsy, for her unflagging love and support. I argue that she read my dissertation in draft form more times than did I, and the same is no doubt true for this volume as well. Thank you.

Abbreviations

ACWA	Association of Clean Water Administrators
ADEM	Alabama Department of Environmental Management (AL)
AFDC	Aid for Families with Dependent Children
ARRA	American Recovery and Reinvestment Act (2009)
ASIWPCA	Association of State and Interstate Water Pollution Control Administrators
AWPCA	Alabama Water Pollution Control Authority
BAT	best available technology
BOD	biological oxygen demand
BPJ	best professional judgment
CAA	Clean Air Act
COVID-19	novel coronavirus-19
CSO	combined sewer overflow
CWA	Clean Water Act
CWSRF	Clean Water State Revolving Fund
DEC	Department of Environmental Conservation (NY)
DNR	Department of Natural Resources (GA)
DWSRF	Drinking Water State Revolving Fund
EFAB	Environmental Financial Advisory Board
EIS	environmental impact statement
EPA (USEPA)	US Environmental Protection Agency
EPD	Environmental Protection Division (GA)
ERTA	Economic Recovery Tax Act
FTE	full-time equivalent
FWPCA	Federal Water Pollution Control Act
FY	fiscal year (federal)
GEFA	Georgia Environmental Finance Authority
GO	general obligation (bonds)

GSP	gross state product
HEW	US Department of Health, Education, and Welfare
IUP	Intended Use Plan
LOC	letter of credit
MAG	management advisory group
MARPOL	International Convention for the Prevention of Pollution from Ships
NEPA	National Environmental Policy Act
NIMS	National Information Management System
NMP	National Municipal Policy (List)
NPDES	National Pollution Discharge Elimination System
NYEFC	New York Environmental Finance Corporation
OLS	ordinary least squares
OMB	Office of Management and Budget
OPA	Oil Pollution Act
PHS	US Public Health Service
POTW	publicly-owned treatment works
PPBS	Planning Programming Budgeting System
PPE	personal protective equipment
SDWA	Safe Drinking Water Act
SWRCB	State Water Resources Control Board (CA)
TMDL	total maximum daily load
TRC	Texas Railroad Commission
TRITAC	Trilateral Technical Advisory Committee (CA)
UNCLOS	United Nations Convention on the Law of the Sea
US	United States
WIFIA	Water Infrastructure Finance and Innovation Act
WOTUS	Waters of the US Rule
WPCA	Water Pollution Control Administration
WQA	Water Quality Act
WRCB	Water Resources Control Board (CA)
WRRDA	Water Resources Reform and Development Act
WTW	wastewater treatment works

Introduction

The genesis for this book grew out of a project for which I served as a Graduate Research Assistant during my doctoral studies at Auburn University in the late 1980s and early 1990s. At the time, the Water Quality Act (WQA) of 1987 was new, and many wondered about the efficacy of the shift from categorical grants to a state-administered block grant, while still enforcing the federal regulatory enforcement regime still in place. Under the tutelage and guidance of Professor John G. Heilman and Professor Gerald W. Johnson, co-principal investigators on a grant from the US Geological Survey, we were tasked with studying the initial implementation of this new program.

That early study revealed a program of surprising nuance and complexity. While some states were able to implement the new program from the outset, many states struggled with early implementation. The initial study also revealed (unexpectedly) a great deal of private sector activity in the financial elements of the program, an observation that became the genesis of my doctoral dissertation shortly thereafter. While this early work painted a consistent picture of early state implementation, many questions about program effects were necessarily left unanswered. In the intervening years, I returned fairly regularly to this program through a series of smaller research projects, all the while wishing I could find time to delve more deeply into the nuances of the program, and to examine the degree to which the WQA (and its then-revolutionary funding mechanism, the Clean Water State Revolving Fund [CWSRF] program) have fulfilled their expectations.

The year 2017 marked the thirtieth anniversary of the Water Quality Act. More importantly, unlike the nearly forty years preceding the WQA, the Act remains virtually unchanged from its original form. With the exception of three minor program changes, the WQA operates today as it did in 1988 (the first year of capitalization grants). In addition, although authorization for funding expired in 1994, the CWSRF program, the financial heart of the program, has received congressional appropriations every year since its inception. The longevity of the

1

Act, along with the continued funding stream, provides a unique opportunity to not only examine the impacts of the CWSRF on state decision-making, but also to assess the WQA as a living example of the tenets of "Reagan federalism." As detailed elsewhere in this volume, the WQA stands as perhaps the best exemplar of the underlying philosophy of Reagan's governance model. In many ways, WQA and the CWSRF are the embodiment of the notion of "comparative advantage" federalism in environmental policy. The argument of comparative advantage is a product of a discussion in the US Environmental Protection Agency (EPA) in the early 1980s (EPA 1983), and is consistent with the model of federalism envisioned by the Reagan administration. This provides an excellent vehicle to determine whether the promises of that model have been met.

The success or failure of the WQA and its revolutionary funding model, the CWSRF program, is of no small consequence. The very nature of the WQA served to redefine the relationship between the national government and both state and local governments. Whereas under the previous categorical grant model funding was provided to communities from the national government to communities with a minimal role for states, the block grant model required states to become the preeminent actor in the policy space. Assumption of this role required two important issues to be addressed. First, states would still be subject to the regulatory enforcement regime developed under the Clean Water Act (CWA) of 1972. States that failed to assume primacy for regulatory enforcement would be subjected to direct enforcement by EPA. Second, the revolving loan fund model required a much higher level of financial management than was common in most state environmental or health agencies. The requirement for the revolving fund to be available in perpetuity, coupled with pressure by the federal government to leverage state funds, meant that the CWSRF would require an administrative and financial capacity heretofore not required under a federal law – in many ways, the WQA offers a new approach of "environmental fiscalism." The WQA would thus test the proposition that states were both willing and able to assume responsibility for meeting their water quality needs.

Major Themes of the Book

In his seminal work *Les Miserables*, Victor Hugo used the sewer system in Paris as a metaphor for his commentary on the social order prevalent in Parisian life. The sewer, he suggested, was where all pretense was stripped away, and everything was mixed together and presented on an equal footing. The egalitarian nature of the sewer represented something desirable, even as he described the sewer as "fetid streams of subterranean slime which the pavement hides from you..." (Hugo [1861] 1931, 1054). The sewer was described as something that was at once both desirable and undesirable; on the one hand, sewer helped prevent the spread of disease in the city, but the sewer also represented a source of valuable fertilizer that was dumped, without regard for its economic value, into the sea. By the twentieth century scientists understood much better the links between sewage and public health. Hugo's lament regarding the economic value of sewage was transformed into a broader concern for the health and safety of the public. Hugo also spoke, at least in the abstract, of the spread of sewage when he described the ways in which sewage flowed from the sewer system of Paris into rivers and streams, ultimately to flow into the ocean. Although Hugo did not speak directly to the health and environmental effects of downstream sewage, twentieth-century scientists better understood the mechanisms of the spread of disease, and realized that sewage treatment was a key element in both the battle against disease as well as environmental degradation.

The Water Quality Act is the most recent mechanism through which the US government seeks to control water pollution, with a particular focus on municipal sewage. The history of water pollution policy in the United States is one of incremental change, and often caught in the ongoing discussion about federalism in the United States. In many ways, federalism is the key concept in the Water Quality Act, in that the legislation served to redefine the roles of the national government and the states in water quality. To understand federalism and the tension between the national and state governments is to understand the water quality arena, especially in terms of point-source pollution. As was the case with the previous water quality legislation, the WQA was clearly focused on point-source water pollution; nonpoint pollution was something of an afterthought (see Adler, Landman, and Cameron 1993). The history of water quality in the United States is one of the definition, and redefinition, of the roles of the national government and the states (and, ultimately, local governments as well). Thus, the impacts of the debate about federalism becomes one of the major themes of this book.

Water quality has proven to be something of a flash point in American politics. On one hand, it is clear that citizens demand clean water. These demands are communicated to the national government, as well as state and local governments. On the other hand, wastewater treatment is expensive, and the groups that bear the brunt of the cost are not always those that enjoy directly the benefits associated with that cost. A city that spends millions of dollars to treat its wastewater that flows into a river does not receive nearly as much benefit as those living downriver. Rivers do not begin and end in political jurisdictions, and as was the case with early efforts to control water pollution, states were often content to let water pollution become "somebody else's problem." Federal policy since 1948 has been aimed squarely at attempts to prevent the spread of pollution and, since the 1960s, at imposing a national standard for water quality.

The second major theme of this work is the role of state choice in water quality policy. The role of states has ebbed and flowed in this policy area, as notions of federalism have changed over time. When the national government first entered the water quality policy arena in the late 1940s, states clearly held the preeminent role. Federal involvement in water quality was limited to a modest funding effort and for the federal courts to serve as means of dispute resolution regarding interstate pollution claims. The trend for the next four decades was an ever-expanding role for the national government in water quality, mostly at the expense of state and local policy control. The national government became the preeminent actor in water quality, in that it not only provided the lion's share of funding for water quality, but also imposed increasingly strict regulatory requirements on states and municipalities. Moreover, the funding was largely controlled by the national government; states had little role in determining which projects in the state were funded. The Water Quality Act in 1987 brought about a sea change in the policy arena. Literally overnight, states were now given a significantly greater role in the water quality arena, and were now required to assume responsibility for a new, complex, and innovative program. The increased flexibility afforded states under the WQA required states to make a number of significant programmatic decisions, all of which would ultimately determine the degree to which their programs would meet the requirements of the WQA. The consequences of failure were not trivial; states who failed to live up to the goals of the Act would find their authority in the water quality arena significantly curtailed.

The third theme of this volume is the importance of implementation in the policy process. While this is no revelation to students and scholars of public policy, the consequences of implementation decisions are nonetheless a critical component of policy outcomes. The ability of a state to meet its water quality goals is inextricably tied to the quality of its implementation of the WQA. States differ in their capacities and will; but to what degree do these differences result in differences in goals? This question is made somewhat more difficult by an important feature of the water quality policy arena: a lack of standardized and meaningful measures of water quality in the states. Several measures are routinely reported and analyzed, but all are fundamentally flawed to the point that they become questionable, at best, as specific indicators of success or progress. In spite of these limitations, many observers suggest that the net effect of federal and state water quality efforts has been to reduce pollution and raise the quality of the nation's waters.

Organization of the Book

This book is organized into ten chapters. Chapter 1 introduces the study and addresses the research questions that form the basis for this work. The chapter also describes the approach employed in this study – a dual frame that combines a metapolicy frame with a frame of state choice. Chapter 2 expands on the former frame by discussing the tenets of Reagan federalism in some detail and examining the larger forces in play that made the CWSRF model an attractive policy choice to address water quality needs in the states.

Chapters 3 and 4 trace the development of national water quality policy from the founding of the nation through the present. Chapter 3 covers the development of water quality policy from the founding of the nation to 1972, as a means to provide context for 1972 amendments to the Federal Water Pollution Control Act. Chapter 4 begins with the development of the landmark 1972 legislation, discusses the impetus for the Water Quality Act in the 1980s, and describes the development and passage of the WQA in some detail. The chapter concludes with a brief discussion of the water quality policy emplaced from 1987 to the present. This discussion aids our understanding of the underlying structure of the WQA and illustrates the largely incremental nature of water quality policy.

Chapter 5 presents the specifics of the WQA, explains the various mechanisms of the Act, and discusses the changes from the Clean Water Act. The chapter focuses on the specifics of the revolving loan fund model as envisioned in the WQA. The loan process, from capitalization grant to recipient loans, is discussed in detail. It also discuss the roles of the states, the national government, and recipient communities. This discussion is framed in a broader conversation of Reagan federalism and states' rights to provide a backdrop for both state and federal action. The chapter also touches on the expressions of congressional intent embedded in the program, which in turn provide a basis by which to examine the outcomes of the Act.

The next section of the book develops and tests a model of state policy choice under the WQA. In Chapter 6, we address some key concepts related to the initial design, development, and implementation of the CWSRF program at the state level. These concepts serve to provide the basis of the analyses in Chapters 7–9. The chapter also presents a model of state choice to explain the distribution of CWSRF program resources to different categories of applicants, including communities with significant environmental need, small communities, financially at-risk (hardship) communities, and to meet nonpoint pollution needs in the state. The model incorporates variables commonly found in the water quality literature, as well as program-specific data from both EPA and from previous studies of the CWSRF program.

Chapter 7 addresses the question of the early implementation of the CWSRF program. Drawing on surveys of state program coordinators conducted in 1990, 1994, 1996, and 2020, The chapter examinees the degree to which states were ready, willing, and able to assume responsibility for the CWSRF program. It also examines a set of implementation choices made by states, all of which helped to determine the nature of the program in the state. Many of those early choices are compared to current state practice, as determined by the results of the program survey conducted in 2020.

Chapter 8 focuses attention specifically on state implementation choices, especially in the early years of the CWSRF program, through a series of four state-level case studies. Each of the case studies develops a more complete picture of the conditions in the state that served to shape the CWSRF program and its implementation. By comparing the different cases, we can place program outcomes in context, and more fully understand both the strengths and limitations of the CWSRF model as a means to fund water quality needs. We also touch on challenges faced by states during implementation.

Chapter 9 offers a detailed national analysis of the distribution of program resources through the WQA and the CWSRF. Earlier work (see Morris 1997; Heilman and Johnson 1991) noted differences in program distribution patterns in the early years of program implementation, but have those patterns persisted over time? What factors explain the observed distribution patterns? To answer these questions, this study employs a national dataset of CWSRF activity over the twenty-nine years of data currently available. The dataset is composed of data from EPA, as well as data collected from states, and supplemented by variables commonly employed in comparative state policy studies.

The book concludes with some summary findings and thoughts regarding the efficacy of the WQA, both as an instrument to meet national water quality needs, as well as an enduring example of Reagan federalism. The WQA has had a profound effect on the implementation of water quality policy, and has fundamentally changed the role of both states and the national government, but to what end? Finally, the chapter touches on the future of both the WQA and the ability of the nation, and the states, to meet water quality needs.

1

Setting the Stage

Clean Water, Federal Policy Goals, and State Choice

Clean water is essential to human life. Water quality has been defined in the United States as a policy problem, worthy of public attention and action.[1] The growth of industry, urbanization, and agriculture, along with increases in human population, has placed much of the nation's supply of water in jeopardy. The issue of environmental protection, especially the protection of water quality, is fundamental to the health and welfare of the population.

The need for public action has been justified by arguments that the critical nature of water quality is truly a public problem, in that the effects of water pollution impact large numbers of people. Water pollution emanates from four primary sources: industrial effluent, residential effluent, urban runoff, and agricultural and nonpoint runoff. Water pollution resulting from industrial production is a classic economic externality – an unwanted or unintended by-product of the manufacturing process (Kraft and Vig 1990, 5). In the absence of regulatory control, industrial effluent historically has been discharged into the environment because there was no economic incentive to treat the effluent before discharge. Residential and agricultural effluents were treated, or not treated, in similar fashion. Moreover, the science of water quality was not well understood. Because of the overload on the ecosystem and the failure of the individual and the market to control the discharge of effluents, governments at all levels have been called upon to address water quality issues. This activity involves the creation and use of multiple standards, sanctions, and incentives for the treatment of waste water. The process is further complicated by the context of US federalism, ambiguity, and conflict in public and private roles.

National attention on the issue of water quality began in a tangible fashion in 1948 with the passage of the Federal Water Pollution Control Act (FWPCA). The Water Quality Act (WQA) of 1987, the subject of this book, is the direct descendant of the FWPCA. For the nearly forty years between the passage of the FWPCA and the WQA, the roles for states and the national government have been defined and redefined. Likewise, the salience of water quality as a public problem has risen and fallen; the various policy responses during those forty years speak to the demands, pressures, and possibilities of the times. The 1972 amendments to the FWPCA, commonly referred to as the Clean Water Act (CWA) of 1972,[2] stand as the zenith of federal involvement in water quality; the CWA significantly strengthened the role of the national government at the expense of state authority and discretion. The CWA also ushered in an unprecedented financial commitment on the part of the national government to meet water quality needs in the states.

The policy instruments[3] of the 1970s, while multiple and varied, were also not universally championed by policy makers at both the state and national levels. Thus, a new policy/program was enacted in the 1980s – the Clean Water State Revolving Fund (CWSRF) program. The program, a federally initiated, state-administered loan program, constitutes an experiment in New Federalism – a return of responsibility to the states for a major public program. The evolution and analysis of this new policy instrument constitute the focus of this book. The study thus addresses fundamental issues of how we "do government" in a federal and mixed (public and private) economic system in which both responsibility and authority are divided between the national government and the states.

Over the years a number of policy alternatives and policy instruments have been proposed, enacted, and implemented in an effort to address issues of water quality.[4] The history of water quality policy in the United States has run from private local control to state initiatives and to centralized federal water quality goals, permits, and sanctions, and from individual state and local funding methods to federal grant and loan programs. As successive policy instruments have been emplaced, they have tended to become more complex (Heilman and Johnson 1992).

Prior to 1972, state and local governments were the actors primarily responsible for the funding of wastewater treatment plants. By the late 1960s, however, a substantial backlog of treatment projects had accumulated. Faced with growing needs and the threat of the accelerated degradation of water quality in the nation, Congress amended the FWPCA in 1972 to establish a robust program of federal grants to support the construction of wastewater treatment plants. The sponsors of the 1972 legislation, known as the Clean Water Act, viewed the grants program as a stopgap measure; the intent was to help state and local water authorities reduce the growing backlog of needs in the short term (EPA 1984). This stopgap view was reinforced through further amendments to the legislation passed in 1977 and 1981 that began a phased reduction in the level and kind of federal grant assistance. Even at this point, the intent of Congress was clear: States bore responsibility for clean water, but the national government would provide substantial financial support to meet state needs. The culmination of this stance was the passage of the

WQA in 1987, which was intended to return responsibility for wastewater infrastructure to states.

Why Does National Water Quality Infrastructure Policy Matter?

Policy proposals for water quality infrastructure have been a part of the national agenda since the New Deal in the 1930s. Although the justification for water quality infrastructure has changed in the intervening years, the issue still demands attention at the national, state, and local levels. The combined efforts of government at all levels have resulted in tens of billions of dollars spent on water quality infrastructure in the last century. Why do these expenditures matter? Why is water quality, and water quality infrastructure, important?

There are at least three reasons that help explain the demands for water quality. First, polluted water constitutes a significant health hazard to humans, as well as much of the nonhuman ecosystem. Polluted water contains bacteria, organisms, and organic and inorganic compounds that can cause great harm to nearly all forms of life. Because water is a finite resource that is used and reused many times over, a failure to remove contaminants from the nation's waters degrades those waters to the point they may be unsafe for any future use. The scientific link between water pollution and health was not well understood until the late nineteenth century, and efforts to curtail water pollution were not widely implemented until the twentieth century. As scientific evidence became more clear, calls for efforts to control water pollution became more strident.

Second, wastewater treatment works (WTWs) are expensive. Even a modestly sized treatment plant can cost tens of millions of dollars to design and build. The planning and design phases alone can take years to complete, as engineers, policy makers, planners, and others must consider a wide range of factors. What is the area to be served? What is the geography of the area? Where will the treated effluent be discharged? What will the capacity of the facility be? Where will/can the treatment plant be sited? How will the effluent be collected and transported to the site of the treatment plant? Once a design is completed and approved, a funding source for construction must be secured. The construction process itself can also take years, depending on the size and scale of the facility. WTWs are typically owned and operated by municipal governments, but comparatively few municipal governments have the resources to invest millions of dollars in wastewater infrastructure; needs for clean water must be balanced against both other capital infrastructure needs as well as ongoing operating costs.

Third, because of national requirements for clean water standards, wastewater plants become a critical factor in growth management at the local level.[5] A WTW that has reached its maximum capacity can no longer connect new customers (homes, businesses) to the system, so new growth is effectively halted. The only way for a municipality to grow is to expand current capacity or to build new facilities.[6] When coupled with additional regulatory requirements, such as a need to treat stormwater, the importance of adequate capacity for treatment becomes paramount. When we consider the widely differing capacities to fund infrastructure across the states, some form of assistance from the national government becomes critical. Because states differ in their capacities, growth rates across states also differ. Moreover, water quality needs are driven, in large part, by population (and population growth). The last several decades have seen a population shift from "Rust Belt" states in the Northeast and upper Midwest to the "Sunbelt" states of the South and Southwest. Water quality infrastructure is not portable; this population shift, and the growth that accompanies that shift, is only possible if new treatment capacity is built in the Sunbelt states.

For decades, the national government has imposed regulatory requirements on states, whether in the environmental arena or in other policy arenas. However, rather than simply impose expensive requirements on states, many of these requirements are accompanied by federal funds meant to reduce the burden of compliance for states. Water quality is no exception. With the significant expansion of regulatory requirements for water quality in 1972, Congress accompanied those requirements with significantly increased funding to support the construction of WTWs, investing billions of new dollars in water quality infrastructure. The funding fell short of the amounts needed; the expectation was that states (and municipalities) would shoulder a portion of the resources needed for compliance. The WQA and the CWSRF, the foci of this book, are the mechanisms through which the national government now provides financial assistance to states (and communities) to fund wastewater infrastructure. Continued compliance with water quality standards and the ability to engage in growth management are thus dependent on continued financial assistance from the national government.

Why Study the Water Quality Act?

The WQA thus provides a somewhat unique opportunity to examine a program that embodies the major tenets of Reagan federalism. The longevity of the law – over thirty years – and the fact that the WQA remains virtually unchanged over that period of time allows us to evaluate the efficacy of the legislation, and determine the degree to which the WQA has fulfilled its initial expectations. The experience of the CWSRF model is not without consequence; a very similar model has been enacted to fund both drinking water infrastructure needs as well as transportation needs. The attractiveness of the state-centric model to meet infrastructure needs is clear, but has it met expectations? Has it met the underlying assumptions of the policy model, and has it lived up to the ideological "drivers" of the policy? Above all, how successful has the program been in terms of cleaning the nation's waters?

Although the WQA contained a number of elements and programs, and was largely an update of the CWA, the WQA also contained a policy instrument not seen before in federal environmental legislation: the Clean Water State Revolving Fund. The impact of the CWSRF as a policy instrument in the environmental policy arena is hard to overstate, in that it essentially changed the relationship between the national government, states, and municipal governments. Although the regulatory framework for water quality remained largely unchanged, the ways in which water quality needs were funded were fundamentally altered by the adoption of the new policy instrument. The CWSRF not only modified the element of fiscal responsibility for water quality, but also relocated decision-making authority for clean water from the national government to state governments. The ability (and willingness) of states to meet national water quality goals is thus brought clearly into focus.

In many respects, the WQA is representative of incremental policy making conducted over a period of about forty years. The heart of the WQA is not very different from the Clean Water Act of 1972; indeed, the 1987 amendments left a great deal of the CWA intact. In several important ways, though, the WQA redefined the nature of water quality policy in the United States, and placed renewed emphases on different kinds of water pollution. By 1987 the National Pollution Discharge Elimination System (NPDES) mechanism had proven largely successful, and the national government had dedicated billions of dollars to water quality needs. Although there were still both current and future point-source needs to be met, the regulatory mechanism in place had met its promise. Still, the stated intent of Congress in the WQA was that the new funding program should meet the needs of three specific classes of applicant communities: communities with significant environmental need, small communities, and hardship (or financially "at-risk") communities.[7] Congress also made it clear in the 1987 legislation that states were to place additional emphasis on nonpoint-source pollution. The CWA had also included provisions for construction grants for nonpoint-source pollution, although very little money was ever allocated for this purpose under the CWA. With the passage of the WQA, Congress rededicated the effort to address nonpoint-source pollution.[8]

Finally, a careful examination of the WQA provides an opportunity to examine the variation in state actions to design, implement, and administer the clean water program. Title VI of the WQA employed a very different form of grant from the previous Title II program, shifting from a categorical grant to a block grant.[9]

The new program allowed states, rather than the federal government, to decide which communities received assistance. However, Title VI also required a very specific financial structure – a revolving loan fund. While states thus had more freedom to target resources from the program, they were constrained in how they could manage the funds. Title VI required the use of a revolving loan fund, with a minimum state monetary contribution. States did have authority to set interest rates, but the original Act limited the loan term to twenty years. States could also choose whether to leverage their funds, a decision some states made out of necessity since the federal capitalization grants fell far short of the amount needed by states to meet their compliance needs. The CWSRF grant was thus simultaneously liberating and constraining. The ways in which states navigated between the constraints and freedoms are brought clearly into focus.

A Move to Decentralization

For much of the history of water quality policy in the United States, the trend has been to enact policy instruments that favor centralized (national) solutions to the problem of water quality. In this regard, water quality policy has followed a trend similar to that of many other policy arenas. Federalism has been defined and redefined over the decades, and that change is an integral feature of American federalism. The question of the appropriate model for federal–state–local relationships is of no small importance; Donald Kettl (2020) refers to federalism as *the* central question of American governance.

   The tension between centralization and decentralization in public policy has been a feature of American governance since the Articles of Confederation were adopted by the Second Continental Congress in 1777. In response to the failures of the Articles of Confederation, our current constitution created a more powerful national government, and sought to strike a balance between the authority of the national government and the authority of state governments. The general trend in practice has tended to favor the national government at the expense of state governments. Moreover, the policies in place at any one time tend to cover a range of philosophies, which serves to illustrate the tension inherent in representative government. For example, Richard Nixon was a strong proponent of revenue sharing, which sought to provide nearly unfettered state authority to allocate federal grant dollars, while simultaneously instituting national wage and price controls. Nixon was also a proponent of national water quality policy, but vetoed the 1972 legislation because of concerns over the costs of the legislation. More recently, President Donald Trump illustrated the tension during the novel coronavirus (COVID-19) pandemic when, within a span of days, he declared that states bore responsibility for securing supplies of personal protective equipment (PPE), and then claimed that the national government could dictate to states when to lift shelter-in-place orders (Bowling, Fisk, and Morris 2020).[10]

By the late 1960s the pressures to reduce the power and reach of the national government gained renewed strength. The election of Ronald Reagan in 1980 gave new voice to the groups favoring states' rights; Reagan campaigned heavily on a platform designed to reduce the size, scope, and cost of the national government, and to return programmatic authority and responsibility to states. While the overall success of Reagan's vision is somewhat questionable, he advocated for a more decentralized approach to governance. The Water Quality Act of 1987 is indicative of this approach, particularly in its move away from a categorical grant program to fund wastewater infrastructure. However, wastewater shares a number of characteristics with other pollution media, particularly what is often an interjurisdictional issue: Pollution does not respect geographic or political boundaries, meaning that pollution produced in one jurisdiction can become a problem for people in a different jurisdiction. This factor has been a significant element of arguments for national environmental policy.

Water quality is representative of a wicked problem (Rittel and Webber 1973). Wicked problems are problems that are so pervasive, so large and complex, that they defy policy attempts at amelioration. Wicked problems typically do not lack for proposed solutions, but they do lack for effective solutions. Because wicked problems are so large and so complex, the available solutions also tend to be both complex and expensive, and tend to gravitate to centralized policy instruments. Moreover, the complexity often serves to obscure the underlying causes, leading to policy disagreements about the nature of those underlying causes (see Stone 1988). Because policy models are almost always causal in nature, disagreement over causes leads directly to disagreements about the veracity of proposed solutions. Modern civilization produces an enormous amount of waste, much of which is discharged into waterways. More importantly, the waterways into which this waste is discharged generally do not terminate in the political jurisdiction in which they are discharged, but rather flow through watersheds into other (often many) jurisdictions, and eventually to the world's oceans. This makes water pollution an especially difficult problem: Because the water carries the waste out of sight, it is relatively easy to make the waste "somebody else's problem." Since the waste itself is an economic externality, there is little market incentive to reduce the waste or treat the waste. In addition, untold millions of tons of waste were discharged into the nation's waters well before the environmental effects of water pollution were generally understood. Finally, to eliminate the discharge of waste, while perhaps desirable, is generally considered to be impossible: The cost to reach a so-called zero-discharge standard would be overwhelming. Wastewater policy thus becomes a balancing act of different interests, goals, costs, benefits, and political possibilities.

The sources of these pollutants are generally broken down into two separate categories. Point-source pollution emanates from sewage treatment plants, industrial operations, and large-scale (industrial) animal farming. These sources are referred to as point-source because the source of the pollution can be specifically determined – think of a pipe discharging effluent into a river. The CWA devised a permitting system, the optimistically titled National Pollution Discharge Elimination System, to identify, track, and limit the effluent these operations can discharge into the nation's waters. The permitting process is backed by a regular inspection and reporting regime. Operators found to be in violation of their permit can face a range of sanctions and penalties. The second category, nonpoint-source pollution, emanates not from a single point but from large, diffuse areas, presenting challenges in terms of determining not only the source but also the parties responsible for creating the pollution. There are many sources of nonpoint pollution, including fertilizer runoff from farms or suburban lawns, pet waste, malfunctioning or unserviceable septic systems, and a variety of other sources. Unlike point-source pollution, nonpoint-source dischargers are not subject to a permitting process,[11] and thus become more difficult to address through public policy.

Water pollution, including point-source pollution, cannot be eliminated in modern industrial society, at least at any reasonable economic cost. At best, we can hope to reduce the amount of pollution discharged into our waters such that the waters are able to remain usable for other purposes – environmental policy makes choices, in effect, between incurring environmental costs and incurring economic costs. In spite of the hopes of Congress in 1972, which set a national no-discharge pollution goal for the United States in the CWA, the technology necessary to achieve this goal has not been developed at a price that makes implementation viable.[12] To that end, the goal of national water quality policy has been to limit the damage inflicted on the nation's waters by pollution through the establishment of a regulatory regime that seeks to limit both the type and amount of discharges into the waterways. The CWA expressed this goal through a desire for "fishable and swimmable waters" by 1984. The no-discharge goal, also set in the CWA, was later abandoned, for all intents and purposes as unrealistic (Glicksman and Batzel 2010).

Another important factor in the arena of water quality policy is that water quality outcomes are notoriously difficult to measure. In the abstract, there are two main approaches to pollution control policy. The first approach is to require pollution control technologies to be applied to the source of pollution. In the case of air pollution, these control technologies might include solutions such as the use of unleaded gasoline and the installation of catalytic converters in automobiles, or the installation of "scrubbers" on smokestacks to remove dangerous pollutants. In water pollution, federal policy has generally focused on the use of point-source technologies, such as large-scale wastewater treatment plants, to remove unwanted pollutants from the water. On a smaller scale, septic systems are designed to remove and break down solid waste in the water, and then filter the effluent through the soil.

A second approach to pollution control policy is through the imposition of ambient standards. In this approach, the focus is not on the source of pollution, but the quality of the environment itself. Again using air pollution as an example, an air quality sampling program tracks the quality of the air over time, and "success" is determined by improvements in ambient air quality. Such efforts are relatively expensive, but many states (and the EPA) measure ambient air quality. Similar approaches are employed in the water quality arena as well. However, measuring ambient quality in water has proven somewhat more challenging (and expensive) than is the case in air quality. There are hundreds of thousands of miles of waterways in the United States, and water quality testing is left almost entirely to states. A glance at the publicly available EPA data on water quality testing (EPA 2021c) reveals a wide range of effort on the part of states to measure and report ambient water quality: Some states report that more than 50 percent of their waters are regularly tested, while other states test less than 5 percent of their waterways.

The US approach to water quality policy thus combines both of these mechanisms to control water pollution: a point-source focused policy of permitting that employs technology to reduce pollution in the waters, and a testing regime to determine the amount of pollution in those waters. Because the latter is difficult to implement on a large scale, determining the "success" of the former is challenging. When coupled with a significant amount of nonpoint-source pollution that affects the nation's waterways, it is even more challenging to determine the degree to which the WQA and the CWSRF have been effective.[13] Few would argue that national water quality policy is unnecessary or a failed policy, but the lack of clearly definable outcomes can prove challenging for policy makers and policy implementers alike.

However, few would suggest that clean water is not a worthy national goal. The enduring question, however, is how best to achieve the goal. Municipal waste treatment, as an expression of efforts to control point-source pollution, has led the way in water quality; ever since the scientific discovery of the link between human waste and disease more than a century ago, policy makers have understood the need to reduce the bacterial loads dumped into waterways. Industrial pollution has followed a similar path in terms of the focus on point-source pollution control, although the science has generally been slower to develop. It is against this backdrop that our national water quality policy developed.

For most of the nation's history, water quality was considered a local government problem. Aside from some federal funding assistance in the 1930s (under the auspices of economic recovery and jobs programs), local governments bore responsibility for wastewater, typically cast as a public health issue. It was not until 1948 that the first national water quality legislation, the FWPCA, was enacted, but it took another twenty-three years before the national government asserted its authority and dominance in this policy realm, albeit under the guise of environmental policy rather than public health policy. For much of the period between 1972 and 1987, the national government was the preeminent actor in this policy arena. The Water Quality Act of 1987 was an attempt to return responsibility for water quality to states and local governments, and to reduce the federal role in the policy arena. As we shall see, the WQA offered some promise of this, although it has yet to meet its full potential in this regard.

Still, the WQA stands today as an important, yet significantly neglected, example of the model of federalism touted by Ronald Reagan. Developed and enacted by a Democratic-controlled Congress during Reagan's second term in the presidency, the WQA reflected many of the major tenets of the New Federalism, including a focus on devolution of policy authority to states, a move to block grants, and a lessened fiscal and regulatory role for the national government, and was promoted as a model of a new approach to federalism. While the WQA has not reached its full potential, it has ushered in a new era in water quality policy, and it has realigned state–federal relationships in water quality. To that end, what has the WQA accomplished, and in what ways has it failed to meet its promises?

The Clean Water State Revolving Fund Program

The WQA contained a number of programmatic changes; perhaps the most profound change was the transition from a categorical grant program to fund wastewater treatment works at the local level. First envisioned in the Federal Water Pollution Control Act of 1948,[14] categorical grants (known as construction grants) were the mechanism through which the national government provided infra-structure assistance to communities to help achieve compliance with water quality standards. The intent of Congress was that the construction grants would supplement state and local funding efforts to reduce pollution from municipal wastewater treatment plants. The construction grants program was the major policy instrument through which funding for wastewater treatment was provided by the national government through the 1981 amendments to the CWA.[15]

The Clean Water Act of 1972 had provided significantly increased funding for the construction grants program. While the stated intent of the 1972 amendments was to *assist* municipalities in meeting their water treatment needs, it became clear by the late 1970s that municipalities were using the grants program as a substitute for state and local funding (Heilman and Johnson 1992, 36). Indeed, by 1976 state and local spending for wastewater treatment works had reached its lowest point in over two decades (EPA 1984). Figure 1.1 illustrates the trends in both federal and

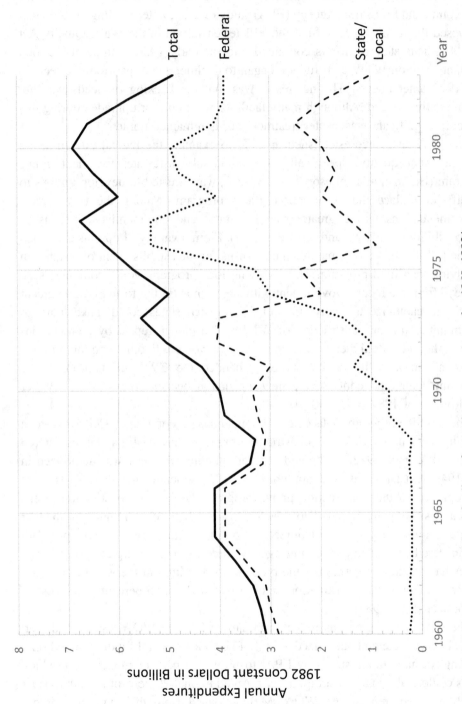

Figure 1.1 Federal, state, and local funding of wastewater treatment works, 1960–1983*

*This figure is adapted from EPA (1984, 3-2).

state efforts in water quality funding; this trend was highlighted in a 1984 US Environmental Protection Agency (EPA) study of wastewater funding, a study that proposed the revolving loan fund model later enshrined in the Water Quality Act of 1987. This state of affairs continued into the early 1980s, although the total amount spent on WTW construction began to decline, due in part to the effects of the 1977 amendments. The net result was that the backlog of needs was not significantly reduced. Although many facilities were built or upgraded during this period,[16] significant wastewater treatment needs remained unmet.

The EPA task force established in 1984 to examine the federal government's role in water quality, specifically the municipal wastewater treatment grants program (Heilman and Johnson 1992, 37–38), was directed to develop options to modify or replace the construction grants program. Among the task force's recommendations was a greater role for states and communities in terms of responsibility for WTW funding (EPA 1984). From a variety of options examined by the task force,[17] there emerged a consensus for the establishment of a state-run revolving loan fund program. The revolving loan program would combine "seed money" from the federal government with state funds (in the form of a 20 percent match) to capitalize a revolving fund from which states could make loans to communities for the construction of WTWs. The money repaid by communities back to the fund could then be loaned again (or revolved) to other communities for the same purpose. In 1987 Congress amended the CWA to incorporate the revolving loan fund concept; the new legislation became known as the Water Quality Act of 1987 (P.L. 100–104).

The CWSRF program replaced the national construction grants (CG) program, a funding mechanism for WTW construction in place since 1948, and administered by the EPA since 1972.[18] Although a small amount of money was authorized in the 1948 legislation, it was not until the 1956 amendments that funds were appropriated to the construction grants program. Under the construction grants program, states were required to develop priority lists for communities in their states. Funds were then disbursed through a categorical grant program administered by the implementing agency directly to the recipient community. States had a role in determining the priority list, but little in the way of authority otherwise. The typical construction grant provided about 75 percent of the cost of the project (EPA 1984).

The CWSRF program, as reflected in Title VI of the WQA, established loan funds in each state and authorized a total of six years' federal funding, based on a funding formula administered by EPA, to provide a pool of money from which states could make loans (see Figure 1.2). An additional 20 percent minimum state match was required by the WQA. Money loaned from the fund to recipient communities would be repaid over time,[19] along with interest at a rate determined

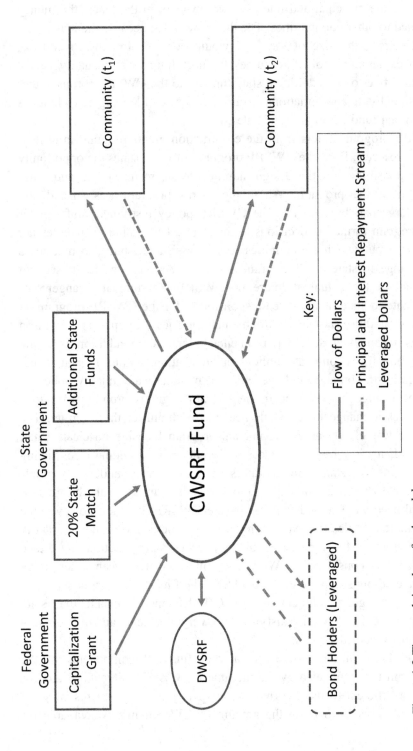

Figure 1.2 The revolving loan fund model

19

by the state. As the principal and interest was repaid from the loans, the money could be loaned to other communities. The theory was that because the loans were repaid with interest, the size of the fund would continue to grow over time, providing an expanding pool of resources to meet future water quality needs. States were also free to add additional state funding to the CWSRF corpus, either through additional state appropriations or through the use of leveraging techniques to raise additional funds through bond sales.

The CWSRF program differs from the construction grants program in at least two important respects. First, the CWSRF program provides states the opportunity to assume responsibility for the design, administration, management, and long-term operation of the program. Where the grants program was shaped and governed in large part by the EPA, the CWSRF policy instrument shifts major policy and program formation decisions to the states. This authority includes not only decisions about which communities receive assistance, but also requires a significant change in state and local funding effort. As analyzed in subsequent chapters, although the actual structure of CWSRF institutional arrangements differs from state to state, many place responsibility for the CWSRF program, at least in part, in traditional state environmental agencies (Heilman and Johnson 1991).[20] This means that state environmental agencies, in addition to being facilitators of federal environmental policy implementation, assume a much greater role in the administration and implementation of national water quality policy.

Second, the change in policy instruments from a grants program to a loan program requires a change in the skills needed to administer the program. The grants program was clearly an environmental program focusing on clean water. The loan mechanism, however, places different and sometimes conflicting financial demands on state administrators. The incentives under the grants program were clearly environmental – applicant communities with the greatest environmental need were served first. Under the CWSRF program, however, the addition of a more complex (and very different) loan scheme meant that in order to meet the requirement for long-term solvency, state administrators must take financial factors into consideration. While still concerned with clean water policy objectives, state administrators must also address the financial components of the CWSRF. The ambiguity inherent in the CWSRF model meant that state administrators were faced with satisfying two very different, and not entirely compatible, policy considerations.

The tension between the environmental and financial components of the CWSRF program is complicated by the fact that the CWSRF allocations are, for many states, insufficient to meet current water quality needs. As a result, some states must find ways to increase the amount of CWSRF money available for loans. Because many states are faced with strong fiscal limitations, they may not be able to invest, for political or financial reasons, additional state resources in the CWSRF pool. One result is that some states have increased their CWSRF funds through financial leveraging of the CWSRF capital pool. Although leveraging schemes vary widely (see Johnson 1995; Holcombe 1992), leveraging generally involves the sale of bonds, backed by money in the CWSRF fund, as a means to increase the current size of the CWSRF resource pool. Leveraging places additional financial demands on the CWSRF, impacts interest rates applied to CWSRF users, and adds to the tensions between the financial and environmental features of the program.

The CWSRF program is thus a policy instrument designed to address point-source water quality improvement needs[21] through the provision of loans for wastewater treatment facility construction or expansion. Because of the nature and structure of the CWSRF program, this particular water quality policy may place environmental and financial priorities of the program in conflict. Due to the financial features of this policy instrument, the CWSRF program provides an opportunity, perhaps an incentive, for the private sector to play an increased role in the institutions created by states to implement the CWSRF program, as well as in the larger clean water policy arena. Likewise, the requirement for the fund to remain solvent in perpetuity places additional pressures on states to ensure that the CWSRF funds are properly managed.

The WQA shifted water quality policy away from federal construction grants to a loan program that would not only allow states to assume primary responsibility for WTW funding, but that would substantially increase state discretion in the management and implementation of the program.[22] National water quality standards, requirements, and enforcement policies remain in effect, yet the discretion given to states means that the policy and program formation roles for state-level actors have fundamentally changed from the grants program.[23] Furthermore, the potential for the revolving loan fund model to meet water quality needs in the future relies much more heavily on the ability of state-level administrators to manage, implement, and fund national policy at the state level (Heilman and Johnson 1991, 196–203).

The CWSRF program is fundamentally unlike any previous national water quality policy instrument enacted in the United States. In addition to providing a great deal of state discretion in addressing water quality issues and policy, the CWSRF program is indicative of a fundamental shift or change in metapolicy, in the relationship between the federal government, state governments, and the private sector. The magnitude of the shift is difficult to underestimate: With passage of the WQA Congress substantially modified a nearly forty-year trend in water quality policy. The Federal Water Pollution Control Act of 1948 created a framework that gave the federal government a role in water quality, a framework (and role) that was expanded and refined through the 1950s and 1960s. By the passage of the Clean Water Act of 1972, the transformation was complete, in that the national government had become the preeminent actor in water quality.[24] The WQA was essentially intended to return responsibility for water quality funding to the states, albeit under the firm control of a federal regulatory and enforcement structure.

The roles for the national government and the states in the area of water quality have been redefined by the convergence of several streams of activity and events of the early 1980s.[25] The net result was a policy instrument called upon to satisfy demands for differing, and sometimes contradictory, goals. If the goal was to shift a major federal program to the states and the private sector, thereby reducing the size and scope of the federal government, the CWSRF program would be a limited success. However, if the goal was to improve and enhance national water quality objectives, the CWSRF program is an experiment that needs to be reviewed and analyzed. Specifically, the CWSRF experiment warrants an analysis of the environmental and financial relationships in the program, the resultant role of the private sector in the formation and implementation of a national public policy, and the impact of these factors on the achievement of national water quality objectives. This study examines the dimensions, effects, and implications of the shift from a categorical grant program to a state-run revolving loan fund program.

One of the underlying assumptions of the WQA was that states were both willing and able to assume financial and administrative responsibility for water quality. The financial arrangements in the revolving fund model were significantly more complex than the grants program it replaced; this meant that states not only had to increase their financial commitment to water quality needs, but also had to accept responsibility for a program that required financial acumen more commonly found in the private sector than in government (and, importantly, in state environmental or health agencies). While some states clearly were prepared (and even eager) to accept this responsibility, it was equally true that other states were not prepared to meet the challenges of the new program (see Morris 1994). It is notable that only eight states were able to accept federal capitalization grants in the first year of the program; it was not until the fourth year of the program that all states had accepted their first capitalization grant.

A second challenge imposed by the WQA was that the funds authorized under the Act were clearly inadequate to meet the water quality needs in the states. Congress and EPA realized this limitation; even when EPA first began to develop the revolving loan model as a policy option, they made clear they understood the funding limitation. In an era of growing budget deficits and decreasing tax revenue, it was evident to most observers that the federal financial effort would be inadequate to meet the goal of state self-sufficiency. To this end, both EPA and Congress proposed the authority for states to leverage their capitalization grants in order to generate additional funds for the program. As we will see, while leveraging certainly does initially create an increased supply of additional funds, it is also the case that leveraged programs come with several costs, including the need for enhanced financial expertise at the state level. More importantly, leveraging also increased the cost of the loans to the borrowers, and presented the opportunity for the fund to lose value over time.[26]

A third challenge to states was the requirement to raise additional state funds. The funding authorization for the CWSRF in the Act was for a total of about $9 billion over eight years; states were required to provide a 20 percent state match for each capitalization grant. Even with leveraging, these funds would not be sufficient to meet then-current water quality needs. There was pressure for states to commit additional funds to water quality; the specter of federal regulation was the "stick" wielded by the national government. But, states were under the same pressures to reduce expenditures and reduce taxes, with the added limitation that most state governments were constitutionally barred from deficit spending. In short, most states did not possess the financial capacity to commit the needed funds to water quality.

As we will see later in this volume, the initial fears of massive under-investment and a resulting "regulatory Armageddon" were unfounded or, at the least, were postponed. Although the original authorization for CWSRF capitalization grants expired in FY 1994, the annual appropriations have continued unabated through FY 2020. From an original federal investment of $9 billion, through FY 2019 Congress has provided roughly $45.6 billion in total for the CWSRF program.[27] Likewise, the WQA authorized Title II funds (also known as construction grants) as a "bridge" between the existing construction grants and the full effect of the CWSRF. The "bridge" funding was authorized through FY 1990, but Title II funds continued to be appropriated in nearly every fiscal year through FY 2013. In short, the program has received more than five times the federal funding originally conceived in the WQA. Since each of these additional capitalization grants required an additional state match, the total investment through FY 2019 is on the order of $54 billion,[28] not including additional funds raised through leveraging or state contributions beyond the 20 percent match requirement. Importantly, however, unmet water quality needs continue to grow, and these needs are spread unevenly across the states (see Figure 1.3).

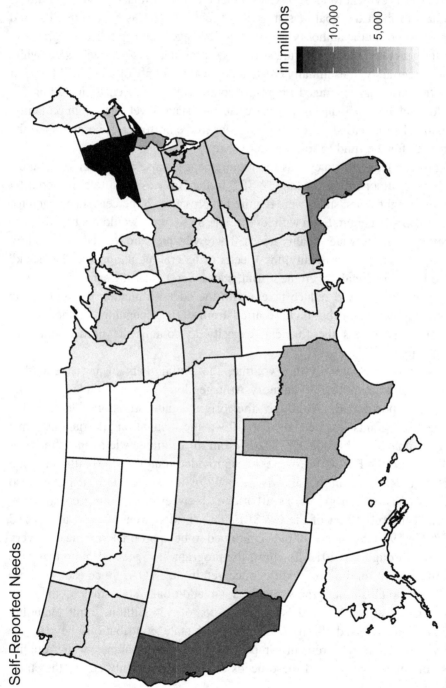

Self-Reported Needs

in millions

10,000

5,000

Figure 1.3 State water quality needs, 2012

A Brief Note Regarding Grant Types

According to Jaroscak, Lawhorn, and Dilger (2020) of the Congressional Research Service (CRS), the CWSRF program is technically not a block grant. They list a total of twenty-three block grant programs in existence, twenty-one of which were funded in FY 2020. The same report defines block grants as a "medium" level of federal administrator discretion, a "medium" level of recipient discretion in the use of funds, and "medium" in terms of the extent of performance conditions (Jaroscak, Lawhorn, and Dilger 2020, 3–4).[29] CRS goes on to define "formula-project categorical grants" as low in federal funding discretion, low in terms of recipient's discretion in the use of funds, and high on performance conditions. The report goes on to state that

[b]lock grant advocates view block grants as a means to increase government efficiency and program effectiveness by redistributing power and accountability through decentralization and partial devolution of decision-making authority from the federal government to state and local governments. They also view them as a means to reduce government expenditures without sacrificing government services.

(Jaroscak, Lawhorn, and Dilger 2020, 2)

While the description of formula-project categorical grants discussed in the CRS report bears some resemblance to the CWSRF grant, this is not an accurate depiction of the CWSRF program. The CWRSF program is formula-based, as block grants tend to be (Wright 1988), and the funds must be used to capitalize state-run loan programs. The WQA also expressly forbids states to use CWSRF funds to award grants to recipient communities. In this sense, the use of CWSRF seems limited. However, states have a great deal of discretion in terms of the recipients of the loans, and for the purposes of the loans made, both in terms of project type and the specific financing type. For example, there are eleven categories of project eligibility and six different categories of financial assistance available, all of which are available at the discretion of each state. The WQA does specify that states must submit both Intended Use Plans (IUPs) and Annual Reports to EPA (see Chapter 5).

David Walker (2000) defines block grants as "an intergovernmental transfer that covers a wide functional terrain and seeks to achieve broad national purposes, while maximizing the discretion of recipient jurisdictions" (Walker 2000, 366 note 7). Discretion in the CWSRF program is not unlimited, but it is arguably broad within the policy arena. States set interest rates and loan terms, decide the allowable purposes for CWSRF loans, and make decisions about recipient communities. Moreover, limited discretion is not antithetical to block grants; as GAO (1995) notes, Congress has consistently added or modified reporting requirements, cost ceilings, and required set-asides to numerous block grant programs.

A review of several well-known texts on intergovernmental relations further supports the case to consider the CWSRF program as a block grant. Deil Wright (1988) describes the characteristics of block grants as more recipient discretion, reduced reporting requirements, awarded based on a formula, and low matching fund requirements. Wright further identifies four forms of intergovernmental aid: loans, grants-in-aid, block grants, and revenue sharing. O'Toole and Christensen (2013) identify four categories of aid: categorical grants, block grants, revenue sharing, and project grants. Block grants are defined by a formula. For O'Toole and Christensen, project grants are competitive, and the size of the award is undetermined until the grant is awarded. Nice (1987) employs a similar typology to O'Toole and Christensen, and notes that project grants are by application, and a federal agency decides whether the grant is awarded, as well as the amount of the grant. In sum, categorical/project grants limit recipient discretion; they are typically competitive; and the award size is undetermined. None of these characteristics describe the CWSRF program.

On the other hand, block grants employ a formula to determine grant size, have fewer restrictions placed on the grantee than is the case with categorical or project grants, and are not awarded competitively. CWSRF grants fit the consensus description of block grants more closely than they fit the consensus description of categorical or project grants. As will be developed later in this volume, the CWSRF program is consistent in structure, purpose, and operation to the description of block grants provided by Jaroscak, Lawhorn, and Dilger (2020), Wright (1988), Walker (2000), and O'Toole and Christensen (2013), among others. In essence the CWSRF program is a block grant in all but name; for the purposes of this book, we will treat the CWSRF program as a block grant.

A comparison of the Temporary Assistance for Needy Families (TANF) block grant with the CWSRF grant program is instructive. The TANF grant is closed-ended, determined by formula and provides states the ability to design a program within a set of parameters specified in the enabling legislation. These same characteristics are found in the CWSRF grant program. Both programs also employ a "carrot and stick" set of incentives. For TANF, states are given greater programmatic flexibility; the "stick" is a loss of entitlements for noncompliance (Lurie 1997). The CWSRF program also provides more programmatic flexibility; the "stick" takes the form of the National Pollution Discharge Elimination System (NPDES), the regulatory framework emplaced to ensure compliance with national standards. One important difference between the two grant programs is that states may carry over funds in the TANF program from year to year, while the WQA requires CWSRF funds to be allocated in a given time frame. However, once the CWSRF funds are paid back to the state in terms of loan repayments, all restrictions regarding future use are removed. A second important difference is that

TANF effectively combines three previous grant programs: Aid for Families with Dependent Children (AFDC), the JOBS program grants, and a small emergency assistance grant program (Lurie 1997). The CWSRF is largely a new, standalone grant program, although in the first three fiscal years of the program, states were allowed to convert Title II funds to the CWSRF program, which effectively consolidated the two grant programs into a single program.

Framing the Analysis: A Two-Level Approach

In order to more fully understand the WQA and its implementation at the state level, this book employs a bi-level analytical frame. The first frame focuses on the broader questions of Reagan federalism and its attendant metapolicy.[30] This book proceeds from the assumption that policy instruments reflect, as suggested by Kingdon (1984), the ways in which policy problems and policy solutions are defined and discussed at the particular time, coupled with the nature of politics at that point in time. Indeed, our discussion of the history of water quality policy in the United States, found in Chapters 3 and 4, employs Kingdon's framework as a means to understand the incremental nature of water quality policy throughout the history of the nation.

The frame of Reagan federalism and the metapolicies in place at the time are reflected directly in the WQA as enacted in 1987. Like many policy instruments it is an imperfect example of its time, in that there are elements of the policy conversation that are not reflected in the WQA. For example, an important element of Reagan's philosophy of governance was the dismantling of the federal regulatory regime, a talking point that has been a prominent element of the Republican Party since this point in time. However, the WQA did little to provide regulatory relief. On the other hand, the WQA removed the categorical grant program delineated in the original Title II (the construction grants program) and replaced it with a new grant program under Title VI (the CWSRF program).[31] The move to the new grant greatly increased state responsibility and authority in wastewater infrastructure funding, and states were incented, both through the WQA and through EPA guidance, to maximize various elements of their programs consistent with the elements of the metapolicy.

An understanding of this broader frame allows us to examine our second frame, state choice, in context. The shifts in metapolicy meant that states were given significant latitude to make programmatic choices under the WQA, even if the overall range of choices was defined and constrained by the metapolicy in place. The focus of the second frame is on a detailed analysis of state choice under the WQA. This will be accomplished through the construction and analysis of a fifty-state model of state policy choice, using a variety of dependent variables that each measures an element of the outcomes possible under the WQA. For example, the WQA specifies the kinds of community states should serve with the program; the models presented will assess the degree to which the CWSRF program has distributed resources to these categories of uses. These models employ a dataset spanning the period, fiscal year 1988 through fiscal year 2016 (the most recent year for which data are complete), giving a dataset containing 1,450 state/year observations, and more than 300 discrete variables. The model for analysis is developed and presented in Chapter 6. Other analyses include a fifty-state analysis of state policy choices made during the design and implementation phases of the program, updated with survey information collected during 2020. The analyses from these surveys are the subject of Chapter 7. Chapter 8 presents a series of case studies that examine the choices made by states during the design, initial implementation, and administration of the CWSRF program. The cases are chosen to span a range of conditions, foci, and circumstances across states. By examining individual state decision processes, we shed additional light on the choices that states made when designing and implementing their programs. Finally, Chapter 9 presents the results of a series of analyses to discern the patterns in state loan distributions during the study period.

Summary

The WQA, and its grant program, the Clean Water State Revolving Loan Fund program, fundamentally changed the nature of infrastructure funding for water quality in the United States. While the overall goal of the WQA did not differ substantially from the CWA, the mechanisms employed fundamentally changed the nature of the relationship between the national government and the states. Rather than the secondary role states played in water quality infrastructure prior to 1987, the WQA thrust states squarely into the central role. Not only were states now responsible for the distribution of program resources to communities within the state, they were now solely responsible for the long-term financial viability of the funding program at the state level. The ability of a state to maintain a financially viable fund, distribute funds in a manner consistent with the intent of Congress, and to meet water quality standards would depend greatly on a combination of state resources and state policy choices.

Were the underlying assumptions of Reagan federalism – that states were ready, willing, able to assume responsibility for the CWSRF program – reasonable, or was it the case that states were not prepared? It is a well-settled question that states differ in their interests and capabilities, but somewhat less is known about the specific policy effects of these differences. If some states lacked the resources or the will to assume responsibility for the CWSRF, what would be the long-term consequences of these shortcomings? Could states make up for initial short-comings over time? What would be the policy consequences for decisions made during the initial design and implementation of the program? These and other questions are at the heart of this book.

The WQA and the CWSRF program are arguably two of the more consequential (and expensive) environmental programs; to date, more than $46 billion has been appropriated for the CWSRF program since FY 1988; states have contributed nearly $138 billion (EPA 2020a) to the CWSRF program during the same period (including dollars raised through leveraging). In spite of this investment, water quality infrastructure needs are significant; EPA reported in its 2012 Clean Watersheds Needs Survey construction needs of $271 billion (EPA 2012), nearly double the total investment in clean water over the past thirty years. With such a huge federal investment and the preeminent role for states in the program, the ability of states to administer the program is brought squarely into focus.

2

Reagan Federalism, States' Rights, and the Revolving Loan Fund Model

The Water Quality Act (WQA) of 1987 was enacted in the second year of Ronald Reagan's second term as president. A long-time advocate for limited national government and a resurgence of states' rights, Reagan championed a model of American federalism and governance that reflected mainstream conservative thought at the time. Though his early political leanings were decidedly liberal (Yager 2006), Reagan moved to a more conservative stance in the 1950s. He first held public office as governor of California, an office he held from 1967 to 1975.

During his service as governor Reagan's conservative record was, perhaps, still being formed. Reagan was a strong advocate for the death penalty (Reagan 1965), but he signed the Therapeutic Abortion Act that ultimately allowed about two million abortions to be performed in California (Cannon 2003). He also signed the Mulford Act that repealed the right of Californians to carry loaded firearms in public (see Coleman 2016). Reagan regularly denounced the welfare state, a theme that would carry over into his years as president. With the advent of revenue sharing in 1972, Reagan became an outspoken critic, arguing that the policy was costly for California. Reagan had also tested the "presidential waters" in 1968, ultimately finishing third at the Republican convention behind Richard Nixon and Nelson Rockefeller. Reagan had a bit more success in the 1976 campaign, although he was denied the Republican nomination in favor of the incumbent, Gerald R. Ford.

By the 1980 presidential campaign, Reagan presented a clear conservative agenda. He campaigned on a platform of limited government, lower taxes, states' rights, and national defense (Lees and Turner 1988; Cannon 2003; Kneeland 1980). In his inaugural address, Reagan famously summed up his philosophy by stating, "[i]n this present crisis, government is not the solution to our problems; government is the problem" (quoted in Murray and Blessing 1993, 80). Reagan's actions during his first term generally followed a clear and consistent pattern. In his first year in office, Reagan pushed through a major revision to the tax code in the form of the Economic Recovery Tax Act of 1981. He also fired more than 11,000 striking air traffic controllers, a move which dealt a significant blow to the union movement (Northrup 1984). Reagan was also a staunch proponent of privatization as an alternative to traditional government provision, production, and delivery of goods and services (see Morris 1999a); the Reagan administration extended and promoted the application of OMB Circular A-76 to direct federal agencies to expand their use of privatization.[1]

The net result of this early activity, and the policy activity that followed, was nothing short of an effort to redefine the relationship between the national government and state governments. Driven by a combination of Reagan's experiences as governor and an increasingly "pure" conservative ideology, Reagan was able to redefine not only the politics stream but also the problem and solution streams (Kingdon 1984). By pressing his agenda and worldview in an alternative politics stream, Reagan successfully reframed both problem definition and the potential solution set. Reagan was largely successful at this in spite of a Congress mostly controlled by Democrats; Reagan won election in 1980 and 1984 with significant majorities in both the popular vote and the Electoral College,[2] and he enjoyed very high approval ratings through much of his presidency. While Reagan certainly did not achieve everything he desired, it is equally clear that he changed the conversation about the role of the national government and the nature of the relationship between the federal and state governments. These redefinitions continue to reverberate in American politics even forty years after Reagan's initial election to the presidency (Kettl 2020). This chapter examines in more detail the nature of Reagan federalism, and how the Water Quality Act of 1987 and its Clean Water State Revolving Fund (CWSRF) program are living examples of the Reagan philosophy of federalism and states' rights.

Metapolicy Change in the 1980s

The 1980s witnessed a pattern of activity in the United States designed to change the nature of the relationships among different levels of government in the US federal system and between governments and the private sector. The new metapolicy of "New Federalism" (Johnson and Heilman 1987; Palmer and Sawhill 1984) or "Reagan Federalism" (Conlan 1988, 3), was driven by a combination of forces or policy streams of activity.[3] One stream driving New Federalism was the issue of decreasing public resources and increasing demands for public services. In a sense, this call for a greater role for the private sector was an important "leg" of the stool upon which Reagan federalism was built. By decreasing the role of the national government and ceding authority and responsibility to the states,[4] new opportunities for the private sector would be realized at both the national and state levels. Reagan's negative view of redistributive programs – social services and subsidy programs – were especially ripe for this criticism. The key to this chain was the willingness of the states to follow the federal lead of retrenchment (Nathan and Doolittle 1987).

The first major pieces of legislation in the economics of New Federalism were the Economic Recovery Tax Act (ERTA) of 1981, and the Tax Equity and Fiscal Responsibility Act of 1982. These Acts provided several incentives to attract private investment to capital-intensive infrastructure projects. Among these incentives were accelerated depreciation schedules, tax-exempt municipal bonds, arbitrage, and a 10 percent tax credit. The net result of these incentives was a brief flurry of privatization activity in wastewater treatment works (WTWs). Traditionally a local government function, this change prompted greater private sector involvement in WTWs, especially from an ownership and operation point of view. Private sector construction and engineering firms and accounting firms developed and promoted privatization strategies. Private sector firms designed, built, operated, and owned WTWs in several municipalities around the nation (see Heilman and Johnson 1992).

The ERTA created market opportunities for private sector involvement in WTW construction (Heilman and Johnson 1992, 44). However, three years later when the focus of federal attention had shifted to budget deficits, the Deficit Reduction Act of 1984 was enacted to limit public-private infrastructure arrangements to long-term contractual agreements (Heilman and Johnson 1992, 43–44). The Tax Reform Act of 1986 removed most of the 1981 incentives[5] and changed the nature of privatization in capital-intensive infrastructure projects (Heilman and Johnson 1992, 44; see also Zimmerman 1987; Watson and Vocino 1990). The salient point is that the activities of the Reagan administration were designed, in part, to enhance the attractiveness of capital-intensive public works projects to the private sector, an interest that remained with the adoption of the CWSRF program.

The changes brought by the Tax Reform Act were designed to reduce the growing federal deficit,[6] an increasingly important public policy issue.[7] Calls for reductions in both federal spending and the federal deficit culminated in the Balanced Budget and Emergency Deficit Control Act of 1985 (P.L. 99–177; more commonly referred to as the Gramm–Rudman–Hollings Act). While the net effect of the Gramm–Rudman–Hollings legislation on the level of debt is questionable,[8] its impact was to raise the consciousness of the public to the issue of the deficit, and to place increased importance on the reduction (or *appearance* of reduction) of federal spending (Thelwell 1990). Thus, Congress was placed in the position of finding new and innovative ways to move programs, particularly domestic programs, off-budget.[9] The CWSRF program was designed to meet this objective, if not the environmental objectives, of the WQA.

A second stream that shaped and motivated New Federalism was an ideologically driven attempt to redefine the role of government in the United States. Based in large part on ideological premises, the 1980s saw a backlash against "big government" at all levels, especially at the federal level. The 1980 elections are often cited as evidence of this mandate,[10] one that called for reductions in both the size and scope of government. Government was to be made smaller, less intrusive, and less proactive in the affairs of both individuals and the private sector. The underlying force behind this movement was a strong belief in the fundamental superiority of the private sector as an agent for the provision, production, and delivery of many goods and services,[11] both public and private (Heritage Foundation 1987; President's Commission on Privatization 1988). The assumption of the fundamental superiority of the private market as an allocation mechanism undergirded much of the policy activity of the 1980s. The incentives provided by the market (profit), coupled with the values attached to a capitalist market (freedom of choice, efficiency, equity, liberty, and self-determination[12]), were viewed as consistent with traditional American values. Much of the public discourse of the period reflects the perceived supremacy of a set of values that allows greater personal freedom, and reduces government interference in the affairs of the individual and the market.[13]

The message was clear – if market opportunities are created, free of unnecessary government interference,[14] the market will flourish, and will provide the public with high-quality goods and services at a lower price than government could provide. Thus, this stream included not only a movement to downsize government, but also to transfer public sector functions to the private sector in hopes of realizing returns in market efficiency,[15] as well as maintaining consistency with a core set of "traditional" American values.

A third stream of activity during this period focused on the demands made on government for action. Although the streams discussed in the preceding paragraphs might indicate less reliance on government, demands for government-provided goods and services continued to increase (Kettl 1988, 7; see also Sharp 1990). The changing relationship between the levels of government complicated this stream. While the federal government sought to reduce its load by devolving programmatic authority and responsibility to the states, it did so in many cases without providing any significant resources for states to act – a concept frequently referred to as unfunded mandates.[16] State governments were also faced with a similar dilemma, in that the problem of decreased resources affected state and local government operations, perhaps to a larger degree than it did for the federal government.[17] Coupled with reduced resources was a lack of support on the part of citizens for increases in the costs of public services and their concomitant tax and user fee increases (Colman 1989, 73–81). The net effect of this policy stream was to shift the burden of demands from the federal government to state and local governments and to the private sector. State governments were increasingly called upon not only to provide goods and services no longer provided by the federal government, but also to take additional responsibility for the implementation of federal policy without the federal government providing adequate resources.[18]

A case in point, and the focus of this book, is the CWSRF program initiated by Congress in 1987. By turning over responsibility to the states, the federal government could realize benefits in all three streams. States would have primary responsibility for the funding of wastewater treatment works (WTWs), thereby removing a substantial fiscal responsibility from the federal budget.[19] Second, the federal government would return programmatic and fiscal authority to the states, thus getting the federal government out of the business of funding WTWs, thereby reducing the role of the federal government in the area of environmental policy. Third, by retaining national standards for water quality and treatment, the federal government could theoretically ensure some level of policy consistency among and between states in achieving the goals of national water policy.[20]

The Trajectory of Federalism

The history of American governance is, in many respects, a story of an evolving relationship between the national government and the state governments. Our current Constitution, ratified in 1791, is a direct response to the limitations of the Articles of Confederation, which itself was one response to the dilemma of federal–state relationships (see Kettl 2020). The practice of federalism has evolved as a result of a combination of court cases, changing conditions, differing ideologies, the definition of new problems, and a multitude of other factors. The result is a patchwork of actions, practices, approaches, and decisions regarding the proper relationship between the national and state governments. There is also a divide between those who treat federalism as a legal issue, and those that treat federalism as a pragmatic, policy-based issue.[21]

Scholars of federalism often disagree about the terms used to describe different federal/state relationships, and also about the periods of time encompassed by each of these terms. While a complete history of federalism is well beyond the scope of this work, we can suggest that the early years of the republic were marked by a form of federalism often referred to as "dual federalism," evolving into "cooperative federalism" (Corwin 1950) in the early decades of the nineteenth century (Zimmerman 2001). Elazar (1962) makes a compelling argument that cooperative federalism is effectively the opposite of dual federalism, in that the latter suggests a clear distinction of functions between the two levels of government. Cooperative federalism suggests that the national government and the state become partners in the same policy realm, and work together to achieve policy goals. In this model, cooperation between the national government and the states is negotiated between (effectively) equal partners (Elazar 1962).

Zimmerman (2001) points to the passage of the amendments to the Federal Water Pollution Control Act (FWPCA) of 1965 as an important turning point in federal–state relationships. The 1965 legislation, covered in some detail in Chapter 3 of this book, is notable for the imposition of national water quality standards for the first time. Up to this point, states bore responsibility for both setting and enforcing water quality standards within their boundaries, but the 1965 Act both set national standards and required states to enforce those standards. This use of preemption[22] by the national government fundamentally changed the nature of federalism. In the water quality policy arena, the 1965 Act is important because it set the tone for both the Clean Water Act of 1972, as well as the Water Quality Act of 1987, in terms of the relationship between the national government and the states. The WQA added an additional change – a switch from categorical grants to block[23] grants – which in turn altered the nature of the fiscal relationship between the national government and the states. The use of grants-in-aid had been considered a reasonable and well-accepted fiscal device in cooperative federalism, but the growing use of grants, coupled with a proliferation of federal mandates and requirements, changes in the nature of society, changes in party politics, and structural changes in the economy, eventually began to undermine their use (Conlan 2006).

A number of scholars have argued that the days of cooperative federalism have passed and have been replaced with a different model. Conlan (2006) refers to the growth of "opportunistic federalism," while Cho and Wright (2001) prefer the term "coercive federalism."[24] The salient feature of coercive federalism is that, rather than serving as a neutral or benign incentive, coercive federalism places more emphasis on federal regulation and the use of active disincentives to discourage unwanted behavior by states. Coercive federalism moves beyond preemption as a tool and, as in the case of the Safe Drinking Water Act of 1974, offers the direct threat of civil and/or criminal penalties for noncompliance (Zimmerman 2001). Knight (2013) highlights a long-standing debate in the courts regarding the balance between cooperative federalism and coercive federalism in lawsuits regarding the EPA's implementation of the Clean Air Act (CAA). Moreover, in the case of the CAA, where one stands depends on where one sits: States tend to think the EPA has been much too coercive in their enforcement of the CAA, while most environmental groups believe the EPA exhibits too much deference in their relationship with states, thus undermining the achievement of national clean air goals.

The net result of this history is a patchwork of approaches and policy instruments that, together, define federal–state relations. In the arena of water quality policy, the trend has been to move more toward what Elazar (1965) might term "coercive cooperation." The evolution of water quality policy has witnessed a growth in the national role in water quality, both in terms of the use of grants-in-aid and the imposition of federal regulatory requirements and the threat of preemption. By Reagan's election to the presidency in 1980, these tensions had become a major element of the political discussion, and certainly of Reagan's philosophy of federalism. Reagan's presidency was marked by various attempts to change the nature of federalism in favor of greater state control, under the umbrella of states' rights. The Water Quality Act is as clear an expression of this philosophy as any policy in force in the twenty-first century, and thus provides a useful vehicle to examine the efficacy of Reagan federalism.

The Elements of Reagan Federalism

The core of Reagan's philosophy of federalism was the normative position that states should be the preeminent actors in public policy. His conception of federalism was driven in part by ideas commonly associated with libertarianism and public choice, coupled with healthy doses of experience and pragmatism. By pushing policy control and responsibility, Reagan could achieve multiple goals: reduce the size of the federal budget, cut federal taxes, provide states with greater latitude and control, and reduce federal regulatory control[25] through the use of several policy innovations. It is also the case that Reagan's notions of federalism rested firmly on two important assumptions: first, states possessed the capacity to assume responsibility for federal programs; and second, that states possessed a desire to assume that responsibility.

State Capacity

The fundamental assumptions of Reagan federalism are that states possess the capacity to assume responsibility for these programs, and that states desire to assume responsibility for national policy.[26] In many ways, the evidence for both of these is, at best, mixed. Moreover, with regard to the willingness to assume responsibility for national programs, there are two sub-elements to the assumption: administrative responsibility and fiscal responsibility. Each of these assumptions is examined in turn.

Administrative Capacity

The Great Society programs of the 1960s represent, perhaps, the zenith of the federal assumption of programmatic responsibility. Programs such as Medicaid and

Medicare, the Food Stamp Act of 1964, Aid for Families with Dependent Children (AFDC), and Title I (also referred to as Chapter I) of the Elementary and Secondary Education Act of 1965 created national programs with national eligibility criteria. States administered these programs with varying degrees of administrative and policy freedom, but all within fairly narrow confines set out by law. The goals were determined by federal law, and states had limited freedom and discretion within these frameworks. Because eligibility criteria, program requirements, and processes were created and monitored by federal agencies, states simply followed guidance provided by those federal agencies. States became "pass-through" entities – in effect, the local source of federal program benefits. Administrative costs were often covered, at least in part, with federal funding, and the overall administrative burden on states (or local governments) was relatively low. In sum, the role of states was to implement national law.

At the same time, these programs also provided significant fiscal resources to provide program benefits to citizens. Significant portions of program costs were covered through federal grants, and the state financial effort required was small in comparison to the national effort. The underlying logic was that the policy problems addressed through these programs were national in scope and could only be addressed effectively through a sustained national effort. Federal funding was often provided through a needs-based formula system: The more need in a state, the larger the proportion of available federal funds a state would receive. This system helped address the wide variation in state fiscal capacity to meet these needs; a wealthy state such as Connecticut had a much greater capacity to address elementary and secondary education needs, for example, than did a poor state such as Mississippi. The federal funding formulae attempted to address these inequities.

By the 1980s, states had amassed several decades of experience in the administration of large national programs. Still, states differed significantly in their administrative capacity. A comprehensive study by Bowman and Kearney (1988) noted significant variation in state administrative capacity. Their study identified four factors of state capacity: staffing and spending, accountability and information management, executive centralization, and representation. Some states, such as Nevada and Mississippi, consistently scored on the low end of the scale, while other states, such as Massachusetts and Ohio, consistently scored on the high end of the scale. Bowman and Kearney (1988) also noted a regional clustering effect using their scores: states in the upper Midwest generally exhibited greater administrative capacity, while states in the South exhibited lower scores. Travis, Morris, and Morris (2004) applied these factors to a study of CWSRF leveraging patterns, and found the measures provided significant explanatory power. In short, the presumption of adequate state administrative capacity was questionable; while some states were likely prepared to assume responsibility for complex federal programs, other states were clearly unprepared.[27] As detailed in other chapters of this book, the CWSRF model, and its attendant emphasis on leveraging, requires a level of administrative and financial sophistication that was simply not present in many states in 1987 or, indeed, for many years following.

Fiscal Capacity

The second element of the capacity argument was that states were able to assume financial responsibility for national policy goals. This assumption was clearly not universal; the WQA, for example, authorized both six years of capitalization grants for state CWSRF (Title VI) funds, as well as three additional years of construction grants (Title II) funds as a "bridge" program. However, even federal policy makers understood that a significant state effort would be required to meet state water quality needs and avoid enforcement actions. To this end, and at the urging of the EPA, the WQA included provisions to allow states to leverage their funds in order to raise additional money with which to capitalize their loan programs. The law placed no limits on leveraging; states were free to leverage as much as possible in the prevailing bond market, or to not leverage at all. In essence, policy makers concluded that states did not have the fiscal capacity to appropriate the necessary funds, but leveraging would allow states to borrow money through private equity markets to raise the required resources. As we will see, even this solution would not be sufficient to allow states to meet their water quality needs. Put another way, the assumption of state fiscal capacity, even enhanced through the use of aggressive leveraging, was likely not a reasonable assumption for all states.

State Desire for Assumption of National Policy Programs

The evidence regarding state desire to assume responsibility for national programs (both fiscally and administratively) is somewhat mixed. Certainly, a portion of the rhetoric present during the Reagan era would suggest that states sought relief from federal requirements. At the same time, Peterson (1984, 226) notes that

No constituency stepped forward to support the president's calls for delegation to the states the basic choices of regulatory policy. Business would seem to be the natural constituency to support regulatory devolution. However, business has generally opposed efforts to substitute state for federal regulation, and frequently has lobbied for federal preemption of state regulatory standards.

The business community clearly saw a greater amount of certainty in a single, national regulatory framework than in a patchwork of conflicting state regulatory policies. Moreover, to concentrate their policy efforts at a single (national) point was certainly easier and more efficient than trying to influence policy across fifty different locations.

A body of scholarly work that examines state willingness to assume primacy in environmental legislation also touches on the issue of state desire to assume responsibility for federal programs. Lester (1986) creates a typology of state environmental behavior based on a level of state commitment to the environment, along with the level of a state's dependency on federal aid. The range of his scale for state commitment is particularly instructive: From a composite score combining a state ranking on pollution potential and the ranking of the level of state environmental control, Lester (1986) reports a range of fifty-three points. The level of dependence on federal financial support for environmental programs also shows wide variation, from a high of 80.8 percent to a low of 24 percent (Lester 1986, 157). Woods (2005) also reports wide variation in state willingness to assume primacy in environmental policy implementation. As of this writing, four states have yet to assume primacy under the provisions of the Clean Water Act.

The states' rights movement was particularly pronounced in the South, a region in which memories of Reconstruction still ran strong (see McKee 2018; Cooper and Knotts 2017; Bullock and Rozell 2018). Reagan's interest in appealing to these voters is evidenced by his decision to speak at the Neshoba County (MS) Fair during the 1980 presidential campaign. In his speech, Reagan touted his ability to appeal to supporters of George Wallace, who ran a strong states' rights-based presidential campaign in 1968. Reagan was widely criticized for appearing at the Neshoba County Fair; the fair was held some 7 miles from Philadelphia, MS, the site of the brutal murders of three civil rights workers in 1964. Seen by others as a continuation of Richard Nixon's "southern strategy" (Maxwell and Shields 2019), Reagan clearly saw a sympathetic audience in the South for the message of states' rights. Although their data are from some years after Reagan left office, Cho and Wright (2001) and Bowling and Wright (1998) suggest that there was growing support for states' rights among public administrators at the state level. If presidential vote and Electoral College totals for Reagan's two presidential elections are any indication, Reagan's focus on states' rights enjoyed support beyond conservative Republicans. We can thus reasonably conclude that there was a desire to reconsider the underpinnings of American federalism.

The Broader Legacy of Reagan Federalism

The effects of Reagan federalism have been examined across a broad range of literatures; these studies began while Reagan was still in office. Conlan (1986) noted a tension in the Reagan administration between Reagan's deeply held beliefs about the devolution of authority to states and his desire to accomplish primary policy goals. For example, Conlan points out that the Reagan administration attempted to force changes in affirmative action plans instituted by local governments, but in doing so was attempting to remove discretion and authority from states and communities for the sake of a larger national goal (as defined by his administration). Conlan (1986) concludes that even though Reagan had committed to a redefinition of federalism, in the end Reagan acted much more in the vein as his predecessors – willing to sacrifice federalism for other deeply held policy objectives. In the end, suggests Conlan, Reagan's concept of federalism was more about cuts to the national budget than it was about a redefinition of federalism.

This point was echoed by Caraley and Schlussel (1986), who examined the patterns of federal grant spending during Reagan's first term. Their analysis indicated that the net effect of Reagan's efforts was to shift many federal programs from a series of smaller, often categorical, grants into larger block grants.[28] Although Reagan desired to remove many of these smaller programs entirely, Congress demurred. In 1981, at the height of Reagan's New Federalism initiatives, Congress only agreed to cut one program. It is noteworthy, however, that an inherent benefit to block grants is that they typically are given with fewer "strings" attached. In other words, block grants are generally awarded for use in a large policy arena, and states are free to choose how the funds are to be spent within the confines of the policy arena. Block grants remove the typical hyper-specificity of categorical grants, which in turn does provide states with increased decision authority about how to expend the funds.

Walker (1991) also noted important distinctions between the promise of Reagan's federalism initiatives and actual behavior. For Walker, the Reagan years were simply a continuation of the trend toward a nation-centered federal system. While Walker notes the changes in funding mechanisms during the Reagan years, he also concludes that the drastic changes forecast by some did not occur. Reagan's attempts at deregulation, devolution, reduction of the federal budget, and economic changes (tax cuts) may be viewed as interconnected (Walker 1991), in that they all rely to some extent on one another. The efforts to reduce the regulatory impact of the national government was also seen as something of a failure; Conlan (1988, 201–202), points out that

[i]n regulation as elsewhere, the Reagan administration has left behind a complex but often disappointing legacy when measured against its original objectives . . . its long-term commitment to this goal [of reducing the IGR regulatory impact] proved surprisingly superficial and its deregulatory accomplishments did not extend beyond its first two years in office.

Walker (1991) also noted that the Reagan administration made a concerted effort to reduce support to regional institution-building initiatives. Cuts to various programs to support regional economic development and river basin management were made, along with many others – twelve of thirty-nine federal programs with a regional focus were terminated (Walker 1991, 112). These actions were not consistent with a desire to build state policy and administrative capacity.

Zimmerman (1991, 26) also points to this duality in Reagan's philosophy:

President Ronald Reagan was Janus-faced with respect to his views on domestic policy. His "public" face emphasized New Federalism initiatives defined in general terms as promoting a return of political power from the federal government to the states, local governments, and the people. His "silent" face, however, encouraged additional centralization of political power in several functional areas and prohibited state economic regulation of certain industries. The president vetoed only two preemption bills, one of which was vetoed on nonpreemption grounds.

The one bill vetoed by Reagan on nonpreemption grounds was the Water Quality Act, but the veto was overridden by a strong bipartisan vote in both chambers of Congress. Zimmerman (1991) concludes that the practice of federalism at the end of Reagan's tenure as president was more coercive than it was when he took office in 1981.

In the environmental arena, Lester (1986) noted that states had generally not replaced reduced federal funding with state resources, and that states were hesitant to assume responsibility for national environmental priorities.[29] In this same vein, Davis and Lester (1989) found a significant variation in state commitment to environmental protection. They noted that while some states, such as California, generally enact policies that exceed national standards, other states, such as Mississippi and Alabama, struggle to meet national standards (and even then, only under coercion). Particularly important, argues Lester (1986), are the fourteen states that have low levels of commitment to environmental protection, but a high dependence on federal aid to support environmental policy initiatives. Lester concludes that decentralization and defunding of federal programs in the environmental arena may have an adverse impact on the ability of many states to provide solutions to pressing environmental problems.

Conversely, research conducted in the years following the Reagan administration suggests a slightly different outcome. List and Gerking (2000) examine the effects of the shift to greater state control during the Reagan years and conclude that, in terms of environmental quality and abatement procedures, there is little evidence of a "race to the bottom." By analyzing a fixed-effects model, they find the majority of time effects to be either statistically insignificant or significant (and consistent) with an improvement in environmental quality in the states. These findings are echoed by Millimet (2003), whose research suggests that by the mid-1980s there was evidence of a "race to the top" in terms of the control of emission of sulfur dioxide and nitrogen oxide. Gerlak (2006) noted that in the arena of water policy, the devolution of responsibility to states helped spur the development of collaborative partnerships to address water quality. Some of these groups have experienced significant gains in local water quality (see, e.g., Morris et al. 2013);[30] yet, as Gerlak (2006) notes, collaborative approaches are not a perfect solution, as challenges resulting from policy fragmentation and ecosystem complexity remain.

Ernst (2010) is even stronger in dissent from this view, noting that efforts to restore the Chesapeake Bay watershed have suffered from the application of what he terms "light green" approaches to policy. Ernst makes a strong argument for more direct federal regulatory and enforcement regimes to clean the bay.

Morris (1999a, 1999b, 1997, 1996) and Morris and Heilman (2002) examined the effects of devolution and Reagan federalism on the CWSRF program. This body of work focused specifically on the effects of private sector involvement in the implementation and administration of the CWSRF program.[31] This work contained three important observations regarding the effects of the devolution of responsibility for water quality. First, states were, by and large, not meeting the policy priorities for the distribution of program resources as set forth in the Water Quality Act of 1987. While communities with environmental need were being served by the program (Morris 1997, 1996), small communities and financially at-risk communities were receiving little in the way of assistance. Funding for nonpoint-source projects, another priority of the WQA, was virtually nonexistent in the early years of the CWSRF program.[32] Morris and Heilman (2002) noted that this finding was consistent with the intent of Reagan federalism to devolve these decisions to states.

Second, the distributional patterns noted in the CWSRF program could be attributed, in large part, to the degree to which states involved private sector actors into the management and operation of the CWSRF program (Morris 1997). The more a state involved private sector actors in the program, the more the loan distribution patterns favored larger, financially solvent communities (and the fewer resources went to small and financially at-risk communities). Morris (1997) concluded that states that were unprepared to assume responsibility for the financial complexities of the CWSRF were more likely to employ private sector consultants, and that these consultants were generally more concerned with the financial solvency of the state's program than meeting the policy intent of Congress.[33] States with significant private sector involvement tended to make distribution decisions more in line with those a bank might make. In short, when a state ceded decision authority over public resources to the private sector, it made a difference in terms of the kinds of communities that received those resources. In many ways, this finding could be cast as an inevitable result of Reagan's larger philosophy of government. Reagan's belief in the superiority of the private sector as an institution to distribute goods and services was enhanced by a program that necessitated many states to seek expertise from the private sector. Perhaps an unintended consequence (Morris 1999b) of the CWSRF model, the ability of states to make these choices (or, perhaps, to be forced into these choices) also served Reagan's desire to empower states to structure and administer policy without undue federal interference.

Third, the propensity of a state to involve private sector actors in the administration of the CWSRF program was a direct function of state administrative capacity. States who reported that their initial program resources (administrative, political, and financial) were sufficient were less likely to rely on private sector expertise (Morris 1996). Heilman and Johnson (1991) observed significant variation in terms of the sufficiency of resources at the time of initial implementation of the CWSRF, and case study data from both Heilman and Johnson (1991) and Morris (1994) underscore the importance of initial capacity. These findings also challenged the underlying assumption of Reagan federalism that states were both willing and able to assume responsibility for the implementation and administration of complex national programs. States clearly differed along this dimension, and those differences just as clearly led to differences in, to paraphrase Harold Lasswell (1936), "who gets what."

Some observers have also noted that the period of history covered by the Reagan years also saw the growth of "coercive federalism" (Kincaid 1990; Cho and Wright 2001; Zimmerman 1991), in which federalism saw a shift from a primary objective of social equity to one of deregulation and the use of federal power to preempt state and local authority to fill regulatory vacuums, coupled with the increased use of unfunded mandates. In short, while the stated goal was to empower state and local governments (Zuckert 1983), the net effect was to empower in some respects, and hamstring in others. Shannon and Kee (1989) and Swartz and Peck (1990) argued for the term "competitive federalism," in which states must fend for themselves to meet needs for basic services. The competition, in effect, was between state and local governments on the one hand, and the national government on the other. The forces that sought to lower federal taxes and federal spending were the same forces that sought the same outcomes in states. Shannon and Kee (1989) argued that the commanding position of the national government in this battle for resources predetermined the "winners" and the "losers." These findings were echoed by Veasey (1988), whose case study of the experiences of Arkansas demonstrated the realization that Reagan federalism meant a much greater fiscal role for states than had been the case in the past.

The CWSRF Program and Reagan Federalism

The move toward a new role for the national government in environmental policy under the Reagan administration actually began in 1981 with the nomination of Anne Gorsuch as EPA Administrator. At the direction of the Reagan administration, Gorsuch undertook efforts to reduce the EPA's budget, cut the number of employees in the agency, and relax the enforcement of agency rules. Although Gorsuch was forced to resign in 1983, her actions while serving as administrator began to stimulate discussions in the agency about the future role of both EPA and its relationships with the states.[34] One of the more insightful examples of this work was a task force report published in mid-1983 that examined directly the future of state/federal relationships in environmental policy, a process mandated by Gorsuch's replacement as EPA Administrator, William Ruckelshaus. Composed of top-level state and EPA water program administrators, the task force was charged with addressing a straightforward question: What kind of relationship should the EPA and state water agencies have? Although the question was straightforward, the answer was not. In essence, the task force was looking past traditional "command and control," regulatory relationships, and toward a redefinition of federal/state relationships. The report concluded that EPA should strive to develop a relationship based on the concept of comparative advantage: EPA should do what states are not able to do, and leave states to do what they can do best (EPA 1983). The seeds of a comparative advantage approach would become an important element of the task force established a year later to develop alternative approaches to water quality funding.

As detailed in Chapter 4, development of the amendments to the Water Quality Act began in 1984, some three years prior to passage. The report issued by the EPA in that year illustrated clearly the replacement effect of the construction grants program on state funding effort, and EPA realized that a continuation of the construction grants program would likely meet with stiff resistance. More to the point, decreasing funding levels for the Title II program threatened the viability of the national water quality program. Taken together, it was clear that the construction grants program could not survive much longer.

The Executive Summary of the 1984 report states the issue succinctly (EPA 1984, viii):

Before 1972, States and localities undertook most of the financial responsibility for wastewater treatment. This was the trend of the times, and the designers of the Federal Water Pollution Control Act Amendments of 1972 intended that it continue. The Federal dollars infused via the new Construction Grants Program were meant to help localities meet Federal water quality standards quickly. States and localities were then to resume control. As the program evolved, its focus changed. Billions in Federal construction grants were spent to meet urgent national needs for wastewater collection and treatment. The role of the Federal government expanded rapidly, while State and local funding of publicly owned treatment works dropped to the minimum level required.

Today, the program is changing again. More and more responsibility for management and funding is being returned to the States. Over the next 5 to 15 years, managers will be increasingly concerned with operation, maintenance, rehabilitation and expansion of existing facilities, rather than mainly with meeting "core treatment needs." What is the best way to deliver limited Federal funds in upcoming years so that facilities will be completed, water quality will be protected, and States and localities will receive incentives for becoming self-sufficient in the long-term?

In addition, the areas of consensus detailed in the report could serve as a roadmap for Reagan federalism. Among other areas, the report specifically lists the following areas of agreement (EPA 1984, ix):

- A gradual transition from Federal responsibility to State and local self-sufficiency should be pursued.
- While Federal aid continues, its level should be certain and States should be able to deploy funds flexibly.
- Delegation of Construction Grants Program responsibilities to the States has been successful and should continue.
- Funds should be distributed equitably to meet core treatment needs <u>first</u>, and States should be given the flexibility to address project affordability issues. [Emphasis in original]

The emerging picture was clear: Any replacement for the construction grants program must include an emphasis on state control and flexibility – in other words, let the states do what they can do best, and let EPA do what they can do best.

However, not all of the areas of consensus were entirely in line with the underlying tenets of Reagan federalism. When it came to issues of compliance and enforcement, the working group indicated a strong bias toward the existing regulatory regime (EPA 1984, ix):

- Major changes in standards and compliance deadlines and efforts are counterproductive.
- Federal funding strategies must promote compliance and be supplemented by strong enforcement actions.

In short, the very areas of consensus among the working group reflected directly the same tensions about federalism in the broader context: States' rights were important, but so were national policy goals. In many ways, the assumptions of the working group were consistent with Elazar's (1965) notion of "coercive cooperation." States were clearly to have a more active role in program design, implementation, and administration, but only as long as they also met the federal regulatory requirements. There seemed to be little appetite within the working group to weaken or revise the regulatory framework already in place.

The proposed solution, a state-administered revolving loan fund program, was essentially a block grant program designed to replace the existing categorical grant program. This idea was not radically new; a clear trend during Reagan's first term had been to consolidate categorical grants into block grants (Conlan 1986).[35] What *was* different, however, was the complexity of the proposed program. A few states, such as New York and Pennsylvania, had created state-run and state-funded infrastructure banks that operated on a loan basis, but the concept was new to most states. Whereas the construction grants program was a simple transfer of funds from the US Treasury to a community for a specific purpose at a specific time, the proposed revolving loan fund program required all states essentially to create infrastructure banks, and to evaluate applicants for creditworthiness, track loan disbursements, loan payments, set interest rates, and myriad other financial functions. States that chose to leverage their CWSRF funds added a significantly greater level of complexity to their programs. While EPA provided technical assistance and guidance, the burden of design, implementation, and administration fell squarely on states. Not only would states bear the responsibility for maintaining a financially viable program in perpetuity, states were also under the constant watchful eye of EPA regulators and compliance officials to ensure that states met national water quality standards.

The tension between coercion and cooperation was reflected in the WQA in other ways as well. While states were given latitude to make their own funding decisions, the law required states to meet the needs of certain classes of communities. Under the "first use" doctrine, the initial use of federal funds had to be for communities with significant environmental need. This included communities with treatment plants not in compliance with federal water quality standards.[36] However, the law also required states to meet other needs as well. For example, the law specifically mentions that funds should be used to meet the water quality needs of small communities (those with population less than 10,000), and hardship communities (essentially, poor communities or communities facing financial hardship). Other portions of the law required states in the Chesapeake Bay watershed to set aside 2 percent of their funds to meet watershed-specific needs. Similar set-asides were included for the waterways in and around New York Harbor, the Great Lakes, and San Francisco Bay delta region. Still other provisions required some states to meet the needs of Native American reservations.[37] Finally, the law also encouraged (but did not require) states to use CWSRF funds to meet nonpoint pollution needs. This last provision drew specific attention from President Reagan in his veto message; he saw the emphasis on nonpoint pollution as an overreach of federal regulation (Office of the Federal Register 1987):

Furthermore, ... this bill would also establish a federally controlled and directed program to control what is called "non-point" source pollution. This new program threatens to become the ultimate whip hand for Federal regulators. For example, in participating States, if farmers have more run-off from their land than the Environmental Protection Agency decides is right, the Agency will be able to intrude into decisions such as how and where the farmers must plow their fields, what fertilizers they must use, and what kind of cover crops they must plant. To take another example, the Agency will be able to become a major force in local zoning decisions that will determine whether families can do such basic things as build a new home. That is too much power for anyone to have, least of all the Federal government.

Pursuant to his interests in reducing the size of the federal budget, Reagan's veto message also addressed the cost of the legislation. His statement also branded the construction grants program a failure, noting that the cost had ballooned from an initial authorization of $18 billion to appropriations of more than $47 billion. Curiously, his veto statement also conflates the authorizations for both Title II ($9.6 billion; the construction grants program) and the new Title VI program ($8.4 billion; the revolving fund program) by claiming that the total cost of the bill would simply continue the failures of the construction grants program. It is not unreasonable to assume that the underlying issue for Reagan was not the CWSRF program (which indeed reflected his desire for greater state autonomy), but rather the overall effect of the clean water program on the federal budget.

In sum, while states perhaps did have more autonomy than was the case under the Clean Water Act, the changes were incremental rather than revolutionary. The regulatory and enforcement framework remained intact, and the legislation added nonpoint pollution as well (see Klyza and Sousa 2013).[38] The inclusion of requirements to meet the needs of certain classes of communities also limited state discretion. Conversely, states were now free to design their water quality programs as they saw fit, and to distribute program resources to communities without direct federal involvement. In a sense, both coercion and cooperation were enhanced in the WQA. Given that the WQA is largely unchanged since it became law in 1987, its longevity and continuity provide a unique opportunity to determine the degree to which the promise of the WQA has resulted in desirable water quality outcomes, and the degree to which states have performed at the level specified by the legislation.

Conlan (1988) notes that of the four types of intergovernmental regulatory programs (direct orders, cross-cutting requirements, crossover sanctions, and partial preemptions) all four are to be found in environmental policy. More directly, all four are also present within the framework of the WQA. Direct orders are present in the form of sanctions for noncompliance and a failure to meet national standards. Cross-cutting requirements required funded projects to comply with a host of other federal laws.[39] The WQA also contains crossover sanctions in the form of threats to terminate technical assistance and other smaller programs under the WQA if states fail to meet the requirements under Title VI. Finally, the fundamental regulatory framework of the WQA, itself a legacy of the Clean Water Act, threatens preemption if states fail to enforce clean water standards within their jurisdictions. Reagan's attempts at deregulation sought to reduce all four types of regulatory requirements (Conlan 1988), although the regulatory framework contained in the WQA remained much as it had under its predecessors.

Funding Levels and Leveraging

The 1984 EPA working group realized that the fiscal realities of the times meant that the federal funding effort to meet water quality needs would fall well short of the effort required to meet state water quality needs. Likewise, it was also clear that the limitations on the national government for funding were also in place in the states. The working group considered several options, finally agreeing on the use of state-level leveraging as a means to increase available capital.[40] To receive a federal capitalization grant, states had to pledge matching funds totaling 20 percent of the value of the federal grant. Some state budgets were so tight, however, that even to allocate 20 percent from a state's general fund might be impossible. Leveraging would provide states the ability to borrow funds through the sale of bonds to apply toward the required match, using the federal funds as a guarantee for the bonds. States could also employ leveraging as a mechanism to raise additional funds beyond the required 20 percent match; states that employed aggressive leveraging could generate a significant amount over and above the combination of the capitalization grant and required match. For states with significant environmental need, this could be an attractive option to raise funds.

In many respects, the effect of leveraging was to pass responsibility for the financial requirements to the states. Because the initial authorizations fell far short of current needs, the CWSRF program might be considered as an "underfunded" mandate, rather than an unfunded mandate (Kelly 1994). Over the twenty-four years after the expiration of the initial authorization Congress would make an attempt to ameliorate this issue; the CWSRF program has received appropriations in every fiscal year since its inception. The dilemma for states was still present: The WQA required states to meet clean water standards, but adequate funding was not available. Still, both the WQA and EPA guidance made clear that states had choices about how to raise additional funds. The WQA explicitly allowed for leveraging, and EPA guidance strongly encouraged states to leverage (see EPA 1988a; 1988b).

Summary

The net effects of Reagan federalism, as found in the literature, are best described as "mixed." There is certainly evidence that elements of Reagan's philosophy regarding the relationship between the national government and the states was realized, at least in part, as was EPA's concept of "comparative advantage." The general movement away from categorical grants toward block grant programs no doubt led to increased state discretion and authority, and allowed states to make choices regarding how to best spend its federal funds within that policy arena.

States did make different choices and, in some cases, made innovative choices, thus supporting the notion of states as "policy laboratories" (Dror 1968).

Taken as a whole, the Water Quality Act of 1987 embodies the state of Reagan federalism as it existed at the time. Elements of the policy design – the movement away from categorical grants in Title II toward the block grant program in Title VI and the ability of states to make distribution decisions – are certainly consistent with Reagan's underlying philosophy of federalism. At the same time, it is notable that the WQA left intact the enforcement regime created under the Clean Water Act of 1972. States were still required to meet federal water quality standards, but had much greater latitude to decide how to allocate funds to accomplish that goal. The shift from federal funding responsibility to state responsibility is also consistent with Reagan's desire to reduce federal spending. However, the critical point is that the desired cuts in federal aid came without concomitant cuts in requirements, thus leading to a significant reliance on unfunded mandates (Posner 1998).

The reliance on leveraging to meet state needs plays no small role in the seemingly disjointed goals of Reagan federalism, as expressed in the Water Quality Act. Leveraging highlights the importance of the underlying assumptions of Reagan federalism and becomes, in many respects, the crux of the issue of state capacity. The WQA and its requirements forced many states to alter their implementation structures for water policy, and some states shifted primary responsibility from environmental agencies to state financial agencies or infrastructure authorities. Although the effects of these changes in state implementation are discussed in a later chapter, the salient point was that the WQA placed states in a reactive mode, in which many states were faced with implementation of a type with which they had little or no experience. Indeed, it was not until the fourth year of capitalization grants that all states were prepared to accept a capitalization grant. This leads us back to the question at the heart of this research: Were states ready, willing, and able to accept responsibility for water quality policy, and what were the effects of this fundamental policy shift? The following chapters examine these questions in detail.

3

The Foundations of Water Quality Policy
in the United States

Efforts to keep America's waterways clean date back to the founding of the nation in the late eighteenth century. The quest for clean water is thus not new, but the intensity of the efforts, the targets of the policies, and the policy instruments employed have varied greatly over the years. Although the national government has been in the water quality "game" for scores of years, efforts to treat wastewater are much more recent. The history of water quality policy in the United States has run from private local control to state initiatives; to centralized federal water quality standards, permits, and sanctions; from individual state and local funding methods to federal grant and loan programs. As successive policy instruments have been emplaced, they have tended to become more complex (Heilman and Johnson 1992).

The story of water quality policy in the United States is a combination of definition, development, redefinition, and incremental change over the course of more than a century. From its rather humble roots as a maritime trade issue, to a health issue, to economic development, to environmentalism, to states' rights, changes in water quality policy have reflected political changes in the nation, as well as changes in the definition and practice of federalism. In this sense, water quality policy becomes a bellwether for some of the most fundamental issues of American politics. Culminating in the Water Quality Act (WQA) of 1987, the longest-lived expression of national water quality policy in our history, the WQA itself is the logical result of this decades-long stream of policy activity. But, to fully understand the WQA and its effects, it is necessary to more fully understand the politics, context, decisions, and forces that led to passage of the WQA.

This chapter, and the following chapter, traces the development of water quality policy in the United States as a vehicle to place the Water Quality Act of 1987 (our most recent revision of national policy) into perspective. The WQA represents a significant departure in terms of the policy instrument employed, which itself is a reflection of a redefinition of federalism that came to fruition under the Reagan presidency in the 1980s. And, while policy instruments have changed over time, the WQA represents a magnitude of change rarely detected in other policy arenas. Still, the WQA itself can also be thought of a logical step in the incremental history of the national role in water quality. The WQA is very much a product of the policy instruments that preceded it, and an expression of the nature of politics and the dominant political values in place at the time the Act was passed.

A Policy Streams Perspective

In this regard, John Kingdon's (1984) "policy streams model" is useful as a heuristic through which to view not only the Water Quality Act of 1987, but the entire development of the federal role in water quality policy. Kingdon's model has been employed across a vast array of policy arenas and geographical locations (see Rawat and Morris 2019). The utility of the model lies in its ability to account for the context in which policy making takes place, in addition to the specifics of problem definition and the solutions proposed for addressing problems. It is important to note that Kingdon's interest is in agenda setting; that is, how issues reach the policy agenda for consideration by formal policy making actors. However, the elements of Kingdon's model are useful to organize and understand the nature of problem definition and proposed solution within the framework of politics as they exist at that point in time.

Kingdon's (1984) model is founded on the "garbage can" model of organizational decision-making developed by Cohen, March, and Olsen (1972), which itself was an attempt to reconcile the concept of limited rationality with Lindblom's (1959) notion of incrementalism. Cohen, March, and Olsen (1972) conceived of a process of organizational decision-making that can be described as follows. An organization moves forward, carrying with it a garbage can full of prepackaged solutions, which are arranged in the can in random order. As the organization confronts a problem, it reaches into the can and pulls out the solution on top and applies it to the problem. If that solution solves the problem, the organization continues its forward progress. However, if the first solution does not solve the problem, then the next solution in the can is applied, and so forth, until a solution is found to solve the problem. The important element is the random order of the solutions available, and the reliance on the solution closest at hand (the solution at the top of the pile in the can). The emphasis here is not on the "best" solution, but rather the first solution that effectively neutralizes the problem. Building on this randomness of elements, Kingdon (1984) proposes three "streams" of policy activity that exist independently but in proximity to one another, and that change over time. The contents of each stream at any point in time can be thought of as random, in that the order of the elements in each stream is both unrelated to the other two streams, and that order is not imposed or controlled.

The first stream is the problem stream. This stream consists of all the policy issues in the domain that are vying for governmental attention. Problems can emanate from a range of sources, from focusing events, crises, changes in critical indicators, or through the work of interested and engaged policy actors (Kingdon 1984). An important element of this stream is the ways in which problems are defined and cast (see Stone 1988); the terms in which a problem is defined not only determines its suitability for government action, but also influences the engagement of interests and the perceived available solutions. Kingdon (1984) suggests that problems can wax and wane in terms of importance, but there are always problems in the problem stream.

The second stream is the political stream. This stream is an amalgam of the national mood (broad changes in public opinion, prevalent ideology), organized political forces (changes in presidential administration, party control of Congress, or the outcome of bureaucratic "turf battles"), or changes in metapolicy (Dror 1968; see also Johnson and Heilman 1987). In short, the political stream helps determine what kinds of things are possible in policy terms at a given point in time. Kingdon (1984) suggests that some of these elements may change quickly, while others change more slowly. The difference in rates of change means that at any point in time the political stream will contain a different set of elements (or, perhaps, the elements will have different emphases) than at a point of time before or after the current point. The final stream is the solution stream. Available solutions are determined largely by policy communities consisting of policy actors from a broad spectrum of society. The relative influence of the actors within the communities (and even the communities themselves) will vary over time. The point of the policy communities is to develop and "float" proposals within the community in order to build support for those proposals. In this way, policy communities determine the solutions (proposals) available in the stream at any point in time.

The point at which the three streams converge offers an opportunity for policy discussion, and for a policy issue to be included on the action agenda. Kingdon (1984) discusses the ways in which streams can converge; in short, streams can converge because of an overt crisis or event; because of growing recognition of the need for policy action; or through the actions of a policy entrepreneur. A policy entrepreneur is someone who uses their influence, stature, or connections to advocate for policy action. While policy entrepreneurs are most often thought of as engaged members of policy communities (members of advocacy groups, or appointed or elected government officials), policy entrepreneurs can emerge from all walks of life (see Morris 2007). Policy entrepreneurs may or may not be successful, but they apply their resources to work for policy change. In any event, policy change does not occur without a coupling of the streams.

The Policy Streams Model and Incrementalism

In many ways, the history of water quality policy in the United States can be understood through a "policy streams" lens. However, it is also the case that water quality policy in the United States is fundamentally incremental in nature. The development of a national presence in water quality policy was not a sudden occurrence, but rather the result of many small steps and small changes, with each successive policy choice reliant on those that came before it. Significant change is rare; indeed, one of the most significant changes in federal water policy was embodied in the Water Quality Act of 1987 through the development of the revolving loan fund model administered through a federal grant. Outside of this event, the development of water quality policy is very much the result of a series of small policy changes over several decades.

Charles Lindblom (1959) described a means of explaining (and perhaps predicting) policy outcomes based on a combination of limits to human decision-making rationality coupled with the realities of democratic political processes. In short, incremental decisions are preferable in part because decision makers lack the time and information to make fully rational decisions; they thus tend to default to information at hand. This often includes information about previous decisions and the effects of those decisions. The other important element is the need to reach consensus to make any decision. In a democratic setting in which many different values, preferences, and ideologies are present, divisions between decision makers can run deep. Trying to convince enough of those actors to support a radical shift from the status quo, absent any pressing indication to the contrary is, at best, a difficult proposition. Given a range of differences, the likely nexus of agreement is at a point somewhere in the middle of the range. As successive decision points are reached the same constraints apply, which means that each successive acceptable solution is unlikely to deviate significantly from the last solution.

Applying the Policy Streams Model to Water Quality Policy

If the combination of Kingdon's policy streams model and incrementalism are helpful as frameworks through which to view the development of federal clean water policy, we should expect to see a series of incremental policy decisions that can be described as consistent with the prevailing national mood at the time, coupled with a series of largely incremental solutions available. At different points we may be able to detect the presence of a policy entrepreneur; at other times, we may find a focusing event or perhaps a change in critical indicators brought the streams into convergence. Likewise, we should expect to see instances where the streams converged but did not result in a policy decision, but rather affirmed the status quo.[1]

In addition, there is some evidence that, especially in the period from about 1950 to 1970, developments in water quality and air quality legislation followed a parallel path. Davies and Davies (1975) point out that the federal government became involved in water quality in 1948 and air quality in 1955. Water quality legislation was enacted in 1948, 1956, 1961, 1965, 1966, and 1970. Air quality legislation was passed in 1955, 1963, 1965, 1966, 1967, and 1970. The general pattern was that the policy solutions enacted in the water quality arena were then enacted shortly afterward in the air quality arena. In other words, as the federal government's role in water quality expanded, the role in air quality expanded in similar ways (Davies and Davies 1975). The pattern reversed itself with the passage of the Clean Air Act (CAA) of 1970; as we will see, much of the underlying logic of the Federal Water Pollution Control Act (FWPCA) of 1972, including the expanded role for federal regulation and the setting of national goals, were adopted from the CAA enacted two years earlier.

It is also worth noting that some observers of water quality policy have examined water quality policy through a different lens. Rogers (1993) suggests that water quality policy can be classified by the underlying purpose: economic development, coordination, and regulation. These classifications began as discrete events at different times but have since merged into a single policy arena (Rogers 1993). The early years were about settlement and trade (economic development); this category included maritime navigation. This approach was not without controversy, as some believed that this policy arena was the purview of states and not the national government. Chief Justice Marshall's opinion in *Gibbons v. Ogden* (1824) explicitly included navigation as falling under Congress' power to regulate interstate trade (Rogers 1993, 48), thus settling, if not ending, the debate.

The Progressive movement saw expansion of the national government into the conservation movement as a means to regulate economic uses and protect natural areas. It was during this period that the national parks program was established, perhaps the nation's most enduring legacy of a conservation ethic. This era also witnessed enhanced protections for federal lands, historic sites (through the Antiquities Act of 1906), and other similar initiatives. However, even during this period there were counterefforts to develop natural resources, as evidenced by the Raker Act of 1913 which allowed the City of San Francisco to construct the Hetch Hetchy dam in Yosemite National Park.[2] This path led directly to the involvement of the federal government in hydroelectric power under the guise of regulation of navigable waterways; it also proved an effective means to regulate private power generation (Rogers 1993, 49–50).

The New Deal era saw a focus on water resources development and improvement of harbors and ports; it also saw funds from the national government made available for public works projects related to sewage treatment and control (Rogers 1993, 50–51). It was also during this era when national resources planning took hold; these public works projects were considered a vital element of economic planning. Originating under the auspices of the National Industrial Recovery Act of 1933, the creation of a National Planning Board for water resources (Rogers 1993, 51) stands as evidence of this approach, as does the 1933 creation of the Public Works Administration. As an element of the means to prosecute World War II in the early 1940s, central planning efforts forced the national government to address coordination problems between agencies, and led to the creation of an interagency structure to address water control, supply, and navigation (Rogers 1993).

A series of commissions formed in the postwar period laid the foundation for legislative efforts to address water issues. Many commissions focused on water supply, although some addressed public works more broadly. Later commissions tended to be more inclusive in a policy sense (Rogers 1993), but all of these commissions tended to focus on all three areas – economic development, coordination, and regulation. While Rogers (1993) details a second round of commission activity, the commission relevant to this book is the National Commission on Water Quality, formed in 1973 on the heels of the passage of the Clean Water Act[3] in 1972. The committee's charge was to consider the steps needed to meet the "stringent" requirements of the CWA, which mandated "fishable and swimmable" waters by 1977, and zero pollution discharges by 1985 (Rogers 1993). The commission realized the deadlines could not be met, and recommended case-by-case waivers rather than a blanket extension of deadlines for 1977, but recommended a blanket extension for the 1985 requirements (Rogers 1993, 61–62).

While Rogers' (1993) interpretation is useful, it also obscures some important elements of policy activity that help explain the origins of the 1987 legislation that serves as the focus of this research. The remainder of this chapter examines the development of national clean water policy in an historical context, following a temporal outline through 1970; the following chapter addresses the development of water quality policy from 1972 to the present. For the sake of brevity, we will focus on the legislative changes (or legislative attempts) that determined the path of clean water policy. A full story might include a full discussion of state policy actions, court cases, and other related events that set the course for federal involvement in clean water policy. It is through this lens that we can more fully understand the Water Quality Act of 1987 and the reasons the Act contained the provisions it did.

The Early Years of Water Quality Policy

At the founding of the nation with the Articles of Confederation in 1777 (fully ratified in 1781), water quality was arguably far from the minds of policy makers. There are several likely reasons for this. First, science and medicine were not developed enough to understand fully the links between clean water and human health; these links were not really understood for more than a century after the founding. Second, the combination of a small, largely rural population, coupled with effectively unlimited supplies of fresh water, meant that there was no visible water quality problem. Finally, the Articles of Confederation set forth an exceptionally limited national government; the framers of that document simply did not conceive of a national role (much less an actual need) for water quality policy. With the results of the Constitutional Convention in Philadelphia in 1787, the powers of the national government were strengthened considerably. The framers also built into the document the flexibility of the national government to adapt to new responsibilities and problems as they might arise. Even with the ratification of the present constitution in 1791, water quality policy was not a part of the national policy agenda.[4]

The first piece of national legislation to address the quality of water was the Refuse Act of 1899 (see Freeman 1990; Davies and Davies 1975; Eames 1970). The point of the legislation was to address the practice of the dumping of waste from ships into harbors. This waste was not just sewage and bilge water, but also included garbage, packing materials, ballast, and other solid waste. It is important to note that the Refuse Act was focused on preventing hazards to navigation – it was a transportation issue, and any benefits for water quality were simply incidental. The waste thrown into harbors became obstacles for ships to avoid, and waste that sank below the water was nearly impossible to avoid. Preventing the disposal of solid waste into the water became a means to enhance the flow of goods in and out of ports.

However, the Refuse Act did contribute one important concept to the later development of water quality-focused legislation, a concept that dominated the policy choices for many years. As Stern and Mazze (1974, 83) point out, the Refuse Act "...confined federal jurisdiction to territorial navigable waters and their tributaries, including non-navigable waters from which refuse matter may originate, in accordance with Congress' regulatory powers under the Commerce Clause to prevent hazards to navigation." This definition of the role of the national government in water policy became a de facto limitation on the role of the national government, and meant that early attempts at national water quality standards were limited to bodies of water that flowed across state boundaries. However, the Refuse Act never really received long-term congressional or political support and is largely considered to be ineffectual in terms of its intended purpose to prevent the disposal of waste by ships (Poe 1995).[5]

As the first piece of federal legislation with environmental protection overtones, the Refuse Act can be seen as the first "building block" in the environmental legislative structure. Although known as the Refuse Act, the act was actually a continuation of the "Rivers and Harbors" legislative track which had been on the national agenda since 1826 (see Rogers 1993; Appendix 1). Previous "rivers and harbors" legislation had focused on the maintenance and enhancement of navigation on inland and near-shore waterways, and the 1899 legislation, in essence, simply extended this legislative path to address another hazard to navigation. The Refuse Act can thus be thought of as an incremental change from previous Acts, and ultimately added only one lasting element of environmental legislation. As we will see, other legislative initiatives built on this element, and expanded through the addition of other elements.

In terms of water quality protection as we define it today, to the extent there were any policies in force, they were found at the state and local levels. As the science of waterborne illnesses progressed in the early twentieth century, policy makers began to more fully understand the dangers of dumping untreated waste into bodies of water that also served as drinking water sources. One of the first federal agencies involved in the water quality arena was the US Public Health Service, which was authorized in the Public Health Service Act of 1912 (P.L. 78–410) to conduct research into sources of pollution in navigable waters (Davies and Davies 1975). The policy connection between water quality and public health would remain in force until the 1960s. To this end, some states and local governments began to adopt new technologies to treat wastewater before it was released back into the environment. Cities were the first to adopt these technologies (Ridgeway 1970); rural areas were too sparsely populated to make such systems economically viable. Even in rural areas, however, state and local laws sometimes led to requirements for the installation of septic systems, and prevented the discharge of sewage into creeks, streams, and rivers. Still, these activities were considered squarely in the jurisdiction of states, and the national government did not seek to require any standards. However, as noted earlier in this chapter, the New Deal programs of the 1930s did include funds to help build sewer systems and treatment plants as a means of economic development and job creation; environmental and public health benefits were at best a secondary outcome.

During the New Deal period there were several attempts to enact pollution control legislation, generally under the guise of public works employment programs. There were also strong differences of opinion about how to address the policy issue. Some Republican lawmakers wanted to define pollution as a question of water resource development and leave the programs under the control of the

Army Corps of Engineers as part of their "rivers and harbors" mission. Republicans from the Midwest favored a strong program of federal enforcement of federal water quality standards. The Roosevelt administration rejected the idea of federal standards but favored addressing water pollution through a series of regional commissions or authorities, following the TVA model recently emplaced (Ridgeway 1970, 42). Under the auspices of the Public Works and Works Relief program, New Deal expenditures for sewers went from $47 million in 1932 to $257 million in 1938. The net result of this effort was that over half the nation's urban population was served by a municipal sewer system at the end of the period (Ridgeway 1970, 43).

A bill was introduced in the Senate by Augustine Lonergan (D-CT) in 1936 to create federal water quality standards and empower the national government to seek injunctions against polluters. A counterproposal sponsored by Senator Alben Barkley (D-KY) and Representative Fred Vinson (D-KY) proposed to create a water pollution bureau in the Public Health Service. The new entity would investigate pollution reports and distribute grants for sewer construction to municipalities, but could not set standards or carry out any enforcement. Versions of the bill eventually were passed by both the House and the Senate, but the resulting conference bill contained little of substance and was ultimately vetoed by Roosevelt (Ridgeway 1970, 44–45). A bill introduced in 1940 would have turned over pollution responsibility to the Army Corps of Engineers, but the bill died in committee (Ridgeway 1970, 46–47).

The important elements of this activity in the 1930s were the twin notions of a direct federal role in water quality, and the mechanisms through which that role might be enacted. The proposal of a construction grants program for municipal treatment was not out of line with other public works programs of the era and demonstrated a federal commitment to financial assistance to communities for this purpose. Likewise, the proposal to create federal water quality standards, while ultimately unsuccessful, set the stage for later debates and policy options. The politics stream of the era certainly favored federal intervention in the economy, and the willingness to make the national government the de facto policy leader, at least in public works construction. Similarly, the solution stream favored categorical grants and an enhanced regulatory role for the federal government. Although it would take a number of years for these streams to come to fruition, the fundamental ideas behind later water quality policy efforts clearly find their roots in the period between 1899 and 1947.

The Dawn of a New Era: 1948–1972

Direct federal involvement in water pollution began in 1948 with the passage of the Federal Water Pollution Control Act (FWPCA). The FWPCA was a modest bill of even more modest expectations, but it did signal a willingness and interest of the national government to exert control over what had previously been a state and local function (Poe 1995). The primary purpose of the Act was to encourage states to enter into intergovernmental agreements about how to protect interstate water resources through the promulgation of uniform legislation (Stern and Mazze 1974, 83). More importantly, it built on the underlying principle of the Refuse Act of 1899 and limited activity only to interstate waters. The Act also expressly forbade federal intervention in the process unless consent was provided by all states involved (Stern and Mazze 1974, 83). The language made clear that states were the preeminent actors in this policy setting, and that the federal government would only become involved as an arbiter to settle disputes.

The FWPCA also did not provide for any federal authority to set pollution or discharge limits on states, nor did the legislation provide any enforcement power or enforcement mechanism. The role of the national government was, in essence, to encourage states to enact pollution control requirements. The FWPCA also contained other provisions that allowed the federal government to engage in research, water quality surveys, and to investigate occurrences of water pollution. The Act authorized the expenditure of research funds for up to four years (Stern and Mazze 1974). The national role in this case was thus limited to research and development (Poe 1995). The FWPCA also had a clear public health focus; the Act created the first Federal Water Pollution Control Advisory Board, which in turn ceded control of water pollution programs to the Surgeon General of the United States. The Act did empower the Department of Justice to seek injunctions against polluters, although the requirement for state consent to federal action meant that this authority was rarely used (Stern and Mazze 1974). The FWPCA also holds the distinction of being the first piece of legislation to authorize funds for wastewater treatment plants to be built to satisfy a policy demand for clean water, through the mechanism of a federal program (Davies and Davies 1975, 31). While various federal work programs in the 1930s had included funds for wastewater treatment construction, these were built for economic reasons, rather than environmental or public health reasons. This is not to suggest that the facilities built under the auspices of New Deal work programs were not useful or environmentally friendly; they certainly were. However, the policy activity in the 1930s was in response to a different policy demand. The FWPCA is the first policy expression of a demand to address wastewater treatment for its own sake.

Although the FWPCA contained an authorization for the construction of wastewater treatment facilities through a loan program, the amount authorized was relatively modest – $22.5 million (Davies and Davies 1975). President Harry Truman had initially recommended a mixed program of grants and loans, but a shrinking economy coupled with inflation fears following the end of World War II led to a change in the administration's policy recommendation (Davies and Davies 1975, 31). The funds were to be distributed to states, and states were left to work out the details of how to employ the funds received. More importantly, appropriations were modest, ranging from a high of $3 million in 1950 to less than $1 million in 1953 (Ridgeway 1970, 48). Although the bill was set to expire in 1953, it continued in force until 1956. Still, the FWPCA is notable for its attempt to engage the federal government in a policy arena previously left to the control of the states. The elements of the FWPCA were consistent with the political realities of the times: An expansion of the role of the national government, but one that still left nearly all of the authority (and much of the responsibility) for wastewater treatment in the hands of state governments. The national government was limited largely to an advisory role. The modest sums involved were also indicative of the economic contraction that followed the conclusion of World War II. Still, the FWPCA represents a tentative step down the path of greater involvement by the national government.

The Water Pollution Control Act of 1956

If the FWPCA of 1948 was a small step toward greater federal involvement in water quality, the Water Pollution Control Act of 1956 (WPCA56) represents a somewhat larger step. Milazzo (2006) argues that real national water quality policy began with the 1956 legislation. The WPCA56 redefined the role of wastewater in the policy realm; it provided the first tangible federal financial support for wastewater treatment; it created the pattern of federal grants that would endure for more than thirty years; and it greatly expanded the federal government's ability to set water quality standards. This legislation also represents the first indication of an active, identifiable policy entrepreneur in this policy arena.

Development of the WPCA56

John Milazzo (2006) provides a detailed account of the actors behind the development of the WPCA56. In short, Milazzo concludes that the WPCA56 (and subsequent legislative activity) was the result of a series of changes in the legislative focus of the House Subcommittee on Rivers and Harbors by its chairman, John Blatnik (D-MN). While the 1948 legislation treated water pollution as a public health issue, Blatnik saw an opportunity to expand his committee's authority and to expand the federal role, by redefining water pollution as an issue at the intersection of economic growth, water conservation, and distributive politics. In the process, Blatnik brought a new set of participants into the agenda setting process for water quality and formed bipartisan support for a policy that would distribute benefits widely (Milazzo 2006).

Blatnik's approach was firmly rooted in New Deal politics (Milazzo 2006). The federal government had invested heavily in municipal pollution control as part of the New Deal, which in turn brought thousands of jobs to communities across the United States. Wartime austerity brought an end to the spending, as was the case with the funds authorized under the FWPCA, but not the need for pollution abatement. In the postwar years a combination of overall population growth and a shift of population to urban areas created a greater demand for municipal pollution control. This trend was exacerbated by the growth of industry, much of which operated absent the need to capture and treat industrial effluent. Blatnik saw an opportunity to align his committee's interests with labor interests, which would give his proposals even more legislative support.

The redefinition of water quality as part of a larger water supply problem also changed the landscape of water policy. The nation had experienced a series of droughts in the 1950s; coupled with population growth, water supply was already a visible policy problem. A report by Wallace Vawter in the Library of Congress' Legislative Reference Service (Milazzo 2006) suggested that while there was an adequate supply of water, pollution abatement was a less costly supplement to building dams and reservoirs or finding new sources of clean water. The report made the argument that polluted water was wasted water, since it was unsuitable for economic uses. By maintaining the quality of existing sources, money could be saved (Vawter 1955 – in Milazzo 2006, 27, note 20).

The 1956 legislation thus fundamentally changed the landscape of water pollution policy in the United States and laid the foundation for what would become known as the Clean Water Act of 1972. The 1956 Act was relatively weak, particularly in terms of enforcement, but the basic policy mechanisms were brought into law in this legislation. The emphasis on New Deal solutions, including a focus on distributive politics, an increased role for federal government, and a healthy respect for states' rights set the tone for the next thirty years of water quality legislation. The legislation that would follow in the 1960s strengthened the basic policy mechanisms found in the 1956 legislation, each incrementally contributing to a larger and more expansive federal role in water quality. These changes mirrored the changes in federalism evident in other policy arenas in the 1960s. As we will see, the underlying argument for water quality policy changed as well, as did the interests aligned on each side of the policy debate.

While benefits would be distributive the costs of regulation would fall on specific interests, mostly businesses, who opposed any regulation. These interests were joined by state pollution control administrators, who resented any attempts by the national government to intrude into their policy arena and threaten state control (Milazzo 2006). There was no support from an environmental lobby, largely because an environmental lobby did not exist in the mid-1950s. What few groups existed generally backed national efforts but had very little influence in policy discussions and were easily outgunned by better-organized and better-funded industrial interests (Milazzo 2006).

Blatnik's response was to broaden the list of beneficiaries of the program, and therefore broaden support for the new program. Blatnik calculated that cities and counties would rally behind a program that would provide a significant portion of the cost (through the use of construction grants), and this would translate into support for the bill among members of Congress (Milazzo 2006, 31). The use of construction grants was a solution well-known in Washington; the recently enacted interstate highway program employed construction grants to distribute funds to states. This solution was already present in the solution stream and had garnered wide support as an element of the highway program. The use of construction grants allowed for the wide distribution of public resources but allowed Congress to retain tight control of the distribution patterns. In an era when the political stream favored "pork barrel" politics, the use of construction grants was an easy choice.

Although the redefinition of the water policy arena was successful, Blatnik was less successful in using this support to increase enforcement and regulatory requirements. The context of federalism still favored states' rights, and there was little support for increased federal regulation. In effect, states were willing to accept federal money, but unwilling to accept federal regulation. Compromises to build a coalition of support thus weakened the enforcement provisions and left states largely free to set pollution limits (Milazzo 2006, 33).

A related issue was how the federal funds would be distributed. Lawmakers settled on an allocation formula that included both population and per capita income, modeled after a funding formula from the Hill-Burton Hospital Act of 1946 (Milazzo 2006, 33). To make sure that large cities would not dominate the allocation, the proposed legislation also included a provision to set aside half the funds for small cities. The initial definition of "small" cities was 100,000 and under, until it was pointed out that the city of Duluth (in Blatnik's home state) would not qualify; the definition was then changed to 125,000 (see Ridgeway 1970). While this broadened the distribution of program benefits, it also steered half of the program benefits away from large cities, which typically had the most environmental need (Milazzo 2006, 33–34).

There was also resistance to the legislation from the executive branch. Blatnik's subcommittee had worked closely with the US Army Corps of Engineers for years on water control projects, and the Corps understood the bureaucratic benefits of participation in the iron triangle (see Heclo 1978). The Public Health Service (PHS), part of the Department of Health, Education, and Welfare (HEW), could have benefited greatly from the new program, and had an easier case to make than the Corps for control of the new program (Milazzo 2006, 35). In addition, the PHS had been the federal agency with initial responsibility for water quality, dating back to 1912 (Davies and Davies 1975). But the PHS did not seize the opportunity to wrest control of the program, and representatives of HEW testified before the committee as outspoken critics of the new spending (Milazzo 2006). Similar resistance came from the White House, which later blocked attempts (in 1957–1960) to increase funding for the program. More importantly, the president's veto message from the 1960 legislation attempted to cast the pollution problem as a local problem, and thus not worthy of federal intervention (Milazzo 2006, 36). Although Congress had accepted a federal role in water quality, the Eisenhower administration did not view water pollution as a national problem. The struggle for federal control of water quality was not yet complete, but the 1956 legislation represented another incremental step toward a greater federal control.

Elements of the Water Pollution Control Act of 1956

The final legislation included several important provisions. First, the program was designed to attach federal funding to existing state and municipal projects. The federal government would contribute up to 55 percent of the cost, and the state (and/or municipality) would contribute the remaining 45 percent (Poe 1995). The cost-sharing approach ensured that the recipients still made a substantial contribution to the project. However, it is notable that the federal share did not vary by size of the community, an issue that would not be addressed for several decades. However, the inclusion of a set-aside for small communities gave states an incentive to address communities other than the largest in the state. Second, the Act created a "minimal" enforcement action by providing for an "enforcement conference" (Poe 1995). While this enforcement capability broadened and formalized the federal role in water pollution control (Stern and Mazze 1974), the federal government still had no authority over pollution in intrastate waters. If pollution crossed state boundaries, the federal government or a state governor could call for a conference to negotiate clean-up plans. The conference could include federal agencies, state and local officials, and "other interested parties" (Poe 1995). This provision thus limited federal action and involvement only in cases involving interstate waters.

Third, the Act relied on voluntary participation and consensus for its enforcement provisions. As was the case with the first FWPCA, the 1956 legisla-tion did not expand the federal government's role in regulation and required an invitation from the states for federal participation. The states' rights element of the political stream was still very much in force, ultimately to the detriment of national water quality. While some states embraced the need for clean water regulation, other states largely ignored the push for clean water. In the end, reliance on voluntary participation, consensus, and discretionary authority undermined the potential for success (Poe 1995). In the end, however, the Act "...broadened the base of federal-state cooperation in water pollution, intensified research efforts, increased grants, and brought about stiffer enforcement procedures for controlling contamination of interstate waters" (Stern and Mazze 1974, 82).

The bill authorized a total of $500 million over a ten-year period, with a limit of $50 million per year. Funds from the FWPCA were also limited to no more than 30 percent of the project cost for each award (Ridgeway 1970, 50). Between 1957 and 1961 the program awarded more than 2,500 grants worth $213 million, which generated $4.50 in local matching funds for every dollar spent – a total of $1.25 billion spent on water quality during the period (Milazzo 2006, 60). The program's success helped generate enthusiasm for the program and overcome the Eisenhower administration's opposition to the program (Milazzo 2006), although annual appropriations were still challenging. By the end of the Eisenhower administration, though, the continued success of the program resulted in increased authorizations from $30 million to $90 million annually in the 1961 amendments (Milazzo 2006, 60).

In the 1950s most state sewage control programs were located in state departments of health, which were generally controlled by medical doctors. The primary interest of the medical community was in "...breaking the disease chain between drinking water and sewage" (Ridgeway 1970, 53), and there was little enthusiasm for any enforcement activity. However, public health officials were equally reluctant to cede control of sewage programs, since these programs brought both increased budget control authority and organizational importance. As long as the locus of the federal program was in the PHS, these state organizations could keep a strong grip on state water quality programs, but they continually resisted efforts to create state regulatory and enforcement regimes.

In sum, the Federal Water Pollution Control Act of 1956 laid the foundation for the next three decades of water quality policy in the United States. The Act also represents a substantial shift in national policy through the expansion of an existing policy network. The structure of the FWPCA is very much a reflection of the policy streams of the time. The political stream favored states' rights, a resistance to federal regulation, and a reliance on "pork barrel" approaches to public policy. The solution stream included a preference for construction grants, a focus on interstate activities, and distributive solutions. The problem stream is somewhat harder to identify in this case, since there seems to have been little demand for a broader national role in pollution control. Indeed, if Milazzo (2006) is correct in his analysis, the impetus behind the FWPCA was John Blatnik's interest in expanding his subcommittee's power. However, the broader interest in water as an economic commodity played into this decision as well, since it is likely that, absent this broader interest, Blatnik would not have been able to form a coalition large enough to ensure passage of the bill.

The final element of Kingdon's (1984) model clearly present in this case is involvement of a policy entrepreneur. Blatnik was certainly willing to devote both his time and political resources to the measure, and he worked to create the necessary coalition to ensure passage. One might argue that Blatnik represents the prototypical example of a policy entrepreneur within government: he has power in his role as subcommittee chair; he is connected to a variety of interests within and without government; he is more than willing to devote his political capital to the goal; and he successfully marshals arguments and data to support his case. These entrepreneurial characteristics are evident in other congressional sponsors through the 1960s and 1970s.

For all of its elements, the WPCA was still an incremental step beyond previous water quality policies. It relied heavily on the model provided by the 1948 legislation, to the extent that the 1956 legislation's most important accomplishment was to incrementally strengthen the federal regulatory role. The 1956 appropriations bills did fund the WPCA, a feat the 1948 legislation failed to achieve, although both Acts authorized the expenditure of funds for construction grants. Both acts favored states' rights over federal action, and thus limited federal action to interstate waters – federal involvement still required the acquiescence of the states (Davies and Davies 1975). However, the die had been cast, and the appropriations that followed allowed a case to be made for expanding the program because of its economic and environmental successes. The relatively broad distribution of program benefits also helped build support for the program, even if the main impetus for the program was something other than environmental protection. The next decade would witness the development of a viable environmental lobby, the growth of public support for environmental programs, and a significant expansion of the federal role in water quality.

The Water Pollution Control Act of 1961

The dawn of the new decade ushered in a series of changes in government and society, specifically in the political and problem streams. The election of John Kennedy as president after a close election was regarded by many as a sign of new optimism and of a new role for the national government. One of Kennedy's first policy initiatives was to place an American on the moon by the end of the decade – an idea that seemed nearly impossible in 1961. However, the optimism inherent in this initiative carried over into other areas as well, and the 1960s saw a significant expansion of the role of the federal government. One of the early examples of this is the Water Pollution Control Act of 1961. While in many ways another example of the incremental nature of water quality policy, the Act represents the addition of three key concepts in the water quality arena. The first of these is the removal of state permission for federal suits to seek injunctive relief (Stern and Mazze 1974, 82), a change for which Kennedy signaled early support (Davies and Davies 1975, 29). Earlier Acts required the explicit invitation of state governments in order for the national government to become involved in pollution issues, even in interstate waters. Although intrastate waters remained in the purview of states, the 1961 amendments now allowed the Department of Justice to seek injunctive relief without the consent of the affected states. The federal role was thus strengthened – incrementally – but strengthened all the same.

The second change expanded the national government's role into intrastate and coastal waters, areas that to this point had been solely within the purview of state governments. The federal government still had very limited authority to address issues in these waters, but the Act did allow for the use of federal procedures for pollution abatement in all waterways within the nation (Stern and Mazze 1974). The third change was to move control of water pollution from the Surgeon General to the Department of Health, Education, and Welfare. Although the creation of the Environmental Protection Agency (EPA) was still nearly a decade into the future, policy makers were beginning to think of pollution control in terms that went beyond public health. This transition would not be complete for several more years, but the 1961 legislation was the first step in that journey. The 1961 amendments also added funding for the creation of seven regional research laboratories that would assist both states and the national government with pollution control. These facilities, in essence an expansion of the research funds first authorized in the 1948 legislation, served to strengthen the federal government's role in pollution abatement by creating an authoritative "scientific" voice in policy debates. The ascendancy of scientific data in policy debates in the 1960s became an important element in policy debates into the later 1960s and 1970s. Moreover, the research laboratories began to collect and analyze data on the condition of the nation's waters; as the science progressed, both citizens and policy makers began to understand more about the threats posed by water pollution. The Act also saw an increase in the authorization for the construction grants program to $90 million for fiscal year 1962.

In sum, the 1961 amendments left the previous legislation largely intact, but served to strengthen key elements of the authority of the national government in water quality. The expansion of the authority of the Department of Justice, the application of federal procedures for pollution abatement to intrastate waters, and the relocation of the implementing authority from the Surgeon General and the PHS to HEW all served to signal increased federal interest in water quality. The political stream at the time was amenable to this increased national role; indeed, the adoption of a federal role as represented in the 1961 legislation was, for example, largely unthinkable in 1948. Although it did not affect the debate of the

FWPCA of 1961, the publication of Rachel Carson's *Silent Spring* the following year is widely considered to be the birth of the modern environmental movement (see Shapiro 1990). The increased focus on the environment at large would ultimately alter the problem stream as demands for clean water grew in both number and intensity.

The Water Quality Act of 1965

Legislative work on what would become the Water Quality Act of 1965 (WQA65) began in 1963 in the Senate. Edmund Muskie (D-ME), a rising star in the Democratic Party, introduced a bill in committee to update the 1961 legislation. The bill passed through the Senate easily but ran into resistance in the House Subcommittee on Rivers and Harbors. Although the subcommittee chair, John Blatnik (D-MN) favored the bill, the process was delayed by representatives from Louisiana and Florida who were under pressure from oil interests to oppose the bill. Blatnik was eventually able to overcome the pressure, and the bill was reported out of committee and approved by the House. The conference bill was signed by President Lyndon Johnson in 1965 (Ridgeway 1970, 65).

The 1965 legislation continued the path of an enhanced role for the federal government in water quality policy. By the mid-1960s the trends that had begun early in the decade were now fully evident, and these changes in all three policy streams created new opportunities for an expanded federal role. As noted earlier in this chapter, the 1960s witnessed a shift in the political stream that favored a greater role for national responses to policy problems, often at the expense of states' rights. While this trend was by no means absolute or complete, the clear trend was toward a philosophy of the national government as a policy leader. The solution stream reflected this shift as well, as the sorts of solutions found throughout the period relied more heavily on federal regulatory structures to induce state compliance. These kinds of solution were evident in other policy arenas such as civil rights, voting rights, social welfare, and education.

Perhaps the most important difference in the 1965 legislation was a change in the philosophy of federal involvement. While previous legislation focused on water quality as issues of public health and drinking water safety, the 1965 legislation "...demonstrated an increased concern about protecting ecological values and in-stream water uses, such as fishing and swimming" (Poe 1995, 5). This shift in justification for water quality policy is indicative of a change in the problem stream. What had begun six decades ago as an issue of economic protection and the safety of navigation had morphed into an environmental problem. The concern about the environment for its own sake was not feasible even ten years earlier, but the growing concern in the United States about pollution in all media gave rise to an expanded role for the national government in the provision of clean water.

In spite of the change in the problem stream, the overall process was still largely incremental in nature. When the WQA65 was developed, lawmakers once again began with the legislation already in place as the starting point. The new legislation made several changes to the existing policy framework, all of which increased the federal government's role in water quality policy. First, states were now required to designate intended uses for all interstate waters within their jurisdiction (Glicksman and Batzel 2010), and then to develop ambient water quality standards for all interstate waters (Poe 1995). Concomitant with these standards was the development of implementation plans to be filed with the newly created Water Pollution Control Administration (WPCA). These plans were to include the identification of pollution from specific sources (Adler, Landman, and Cameron 1993, 7), a process difficult to achieve given the technology available at the time (Poe 1995). Once the pollution sources were identified and the ambient water quality standards set, states would then carry out the implementation standards to reduce pollution. If states failed to act, then the federal government would impose standards on the states (Stern and Mazze 1974). As with previous legislation, however, these requirements (and the new authority given the national government) still applied only to interstate waters – intrastate waters were still clearly within the domain of the states.

The legislation also contained a plan to consolidate the water pollution function in the Department of Health, Education, and Welfare, and to create a new organization that would report directly to the secretary of HEW (Ridgeway 1970, 63). The formation of the new organization, the WPCA, also signaled a clear intent on the part of Congress to highlight the importance of clean water. By centralizing the administration of water quality policy, Congress identified a responsible federal organization to address water quality. In some ways this was largely a symbolic gesture, in that the WPCA simply centralized authorities and responsibilities that had already been in place in other federal agencies. By creating the WPCA, however, Congress not only signaled the changing landscape of environmental legislation, they also provided a conduit into federal water policy for environmental interests. In this way, Congress could both build and solidify support for the increased federal role through the creation of a new "iron triangle."[6] The shared interests of congressional subcommittees, agencies, and interest groups would prove invaluable for the next decade of water policy activity.

In 1966 Congress again addressed water quality in the form of the Clean Water Restoration Act (Davies and Davies 1975). The Act authorized additional funds for treatment plant construction, but subsequent appropriations were much less than anticipated (Stern and Mazze 1974). Perhaps more important in the legislation was a push to move control of water pollution from HEW to the Department of the Interior. This effort was spearheaded by Stewart Udall, Secretary of the Interior. A former Congressman and influential cabinet member, Udall forged a deal with Senator Muskie and Congressman Blatnik: If they agreed to the move, both could keep their committee posts in the Congress, and two new committees on public works would be formed. Both agreed, and the program was moved to the Department of the Interior. Many PHS officials refused to move with the program, however, resulting in something of a "brain drain" in the program. Infighting between the staff, coupled with benign neglect from Udall, meant that the program was largely ineffective (Ridgeway 1970, 67–69) until the passage of the Clean Water Act in 1972. The act also created a framework for moving control of water pollution from a state-by-state basis to a regional focus in some river basins (Davies and Davies 1975). The act also lifted the dollar cap on individual construction grants, and allowed for an increase in the federal share of construction costs.

The resultant changes to water quality policy were highly incremental in nature. The Act authorized additional funds for treatment plant construction through the mechanism of the construction grants program, but actual appropriations fell well short of the authorized limits (Stern and Mazze 1974). This Act was followed some four years later by the Water Quality Act of 1970 (WQA70). The 1970 legislation contained three important changes. First, the Act "...extended control standards and procedures to oil and hazardous substance discharge from on- and off-shore vessels and facilities" (Stern and Mazze 1974, 83).[7] In the wake of the oil spill off the coast of Santa Barbara, California in 1969, the potential damage to the marine environment from the discharge of oil was a significant topic of policy discussion, and many Americans were horrified at the images of the damage done to the scenic coastline and the marine wildlife in the region (see Rinde 2017). Second, the WQA70 gave control of water pollution abatement to the newly formed US EPA. Created by an executive order of President Richard Nixon, the formation of the EPA was a response to growing demands in the nation for a more effective federal response to what many saw as a growing environmental crisis in the country.[8] In many ways, the creation of the EPA continued the approach taken by Congress in 1961, augmented in 1965 and again in 1966, to raise the visibility of the federal commitment to environmental policy. As an independent agency, the EPA administrator did not report to a department secretary, but directly to the president. Its independent status also meant that the EPA would not be burdened by an existing organizational culture but could develop a culture of its own.[9] Third, the act extended ambient water quality standards to all navigable waters, and required states to develop implementation plans to meet effluent limits (Milazzo 2006, 194). As before, the changes brought about by both the 1966 and 1970 legislation were incremental in nature, and reflective of the continuing trends in the political, problem, and solution streams.

Between 1956 and 1969 the FWPCA (and its amendments) disbursed about $1.2 billion in the form of more than 9,400 projects worth a total of $5.4 billion (Ridgeway 1970, 98). Including the funds spent during the New Deal era and in 1970, the federal government had invested some $9 billion in sewage plants, which together served nearly 92 percent of the nation's population (Ridgeway 1970, 168). Still, the money allocated fell well below that needed to address water quality in the states. A 1969 report by the US General Accounting Office detailed several instances of the misuse of federal funds. While the report was not designed to demonstrate the severity of the water quality problem, it did so very effectively. The report also demonstrated how well the federal program masked the severity of the problem of water quality (Ridgeway 1970, 98–107).

Summary

The early years of water quality policy were thus characterized by a slow, steady transition from a purely state-run function to a growing federal presence in the policy arena (see Table 3.1). Concomitant with this change of roles was a change in problem definition. Much of the early interest in water quality focused on issues relating to the discharge of pollution from ships in harbors, and efforts to control that discharge for the safety of navigation. Efforts to expand the federal role in water quality in the 1930s were targeted through two goals: the need for public works projects as economic stimulus, and benefits to public health. By the late 1940s the policy problem was still defined as public health, but as the political system became more responsive to broader demands for environmental policy in the 1960s, water quality was redefined again as an environmental issue rather than a public health problem.

Although the Federal Water Pollution Control Act of 1948 created a framework for policy that would remain largely in place for nearly forty years, federal focus (and federal spending) for water quality was not immediately forthcoming. From a small amount of money appropriated following the passage of the 1956 amendments, federal funding effort for water quality grew steadily for the next fifteen years. As funding grew, so did the regulatory role for the national government. These changes, though, were highly incremental in nature, and represent a steady, albeit slow, progression toward a greater federal role that would culminate in the passage of the 1972 amendments. More importantly, these changes occurred as the problem, solution, and political streams converged to provide opportunities for policy change. As the national policy environment was changing, so changed water quality.

Table 3.1 *Major elements of early water quality legislation, 1899–1970*

Act	Year	Major Elements
Refuse Act	1899	Established federal control over navigable waters and tributaries to prevent hazards to navigation
Public Health Service Act	1912	Authorized research into sources of pollution in navigable waters; defined water pollution (in part) as a health issue
Federal Water Pollution Control Act	1948	Encouraged states to enter into intergovernmental agreements to control pollution; authorized funds for research, water quality surveys, and to investigate instances of pollution; created the Federal Water Pollution Control Advisory Board; authorized funds for loan program for wastewater needs
Water Pollution Control Act	1956	Created construction grants program; expanded federal government's ability to set water quality standards; broadened justification for pollution control to include economic development and water conservation; created set-asides for small communities
Water Pollution Control Act	1961	Federal suits for injunctive relief could be brought without state permission; allowed for federal action in both interstate and intrastate waters; creation of regional research laboratories; expansion in construction grants program; movement of program from Public Health Service to Department of Health, Education, and Welfare
Water Quality Act	1965	Focus on water pollution as an environmental issue; state requirement for development of ambient water quality standards for all interstate waters; creation of Water Pollution Control Administration; imposition of federal standards for noncompliant states; relocation of program from HEW to Department of Interior
Clean Water Restoration Act	1966	Authorized additional funds for construction grants; increased federal share of construction costs; develop regional plans for water pollution control
Water Quality Act	1970	Included oil and hazardous substance discharges from on- and off-shore vessels; moved wastewater program to EPA; extended water quality standards to all navigable waters; required states to develop implementation plans to meet effluent limits

This pattern thus set the stage for the significant expansion of the federal role that came about through the passage of the 1972 amendments, the more limited programmatic changes that occurred between 1977 and 1981, and the redefinition of the federal role as captured in the 1987 Water Quality Act. In the next chapter, we explore the development and passage of the 1972 amendments, the series of minor adjustments that were adopted prior to 1987, and finally to the process that led to the development and passage of the 1987 legislation.

4

Expansion and Contraction in the Federal Role in Water Quality Policy

The decade of the 1960s witnessed a significant expansion of federal authority across a wide range of policy arenas. Led by Lyndon Johnson's Great Society initiative and efforts to compel states to guarantee civil rights to all citizens, the net result was a multifaceted set of federal policies that had a profound impact on the theory and practice of federalism in the United States (see Wright 1988; Kettl 2020). The new initiatives included policy areas such as health care (Medicaid and Medicare); education (Head Start, the Chapter 1 program, and Pell grants); poverty (Aid for Families with Dependent Children); and economic development and job creation (Economic Development Administration). In addition, legislative initiatives such as the Civil Rights Act of 1964 and the Voting Rights Act of 1965 sought to expand the national government's influence and power in the states in ways that were resisted by some and welcomed by others. Coupled with the increasing demands for government to address an increasingly visible pollution problem (discussed in the previous chapter), the stage was set for a more robust and proactive effort on the part of the national government to address water pollution.

An Expanded Role for the Federal Government: 1972–1986

The solution stream in the 1960s relied heavily on systems thinking. Inspired by Robert McNamara, Secretary of Defense under Presidents Kennedy and Johnson, and his budgeting model, Planning Programming Budgeting System (PPBS), systems thinking was a more holistic approach to policy problems. Although based on a corporate model of decision-making, it was applied in policy circles to both the defense and domestic policy arenas. Ridgeway (1970, 199) also points out that systems thinking was an effective mechanism to open the policy arena to corporations, which in turn gave the private sector a much broader role in policy making and policy implementation. This move toward greater private sector involvement ultimately rose to new heights in the Reagan years, culminating in the revolving loan fund model that became the centerpiece of the Water Quality Act of 1987. Like many other elements of water quality policy the revolving loan fund model was indicative of a trend found in the solution stream decades earlier, and which found new life (and support) in the politics stream in later years. First, however, the federal role in water quality would need to be expanded.

The Clean Water Act of 1972

Formally known as the 1972 amendments to the Water Pollution Control Act, the Clean Water Act of 1972 (CWA)[1] effectively completed the transition from a mix of federal and state control over pollution to one of federal dominance in water quality policy. The CWA was not revolutionary in the sense that it created a new regime for water quality policy; in essence, the CWA was the culmination of policy trends that had been in place since 1948. With that in mind, it is also the case that the CWA represented a new breadth and depth of federal involvement in water quality and the triumph of "command and control" regulatory structures.[2]

The approach to the Clean Water Act was driven by a broader trend in policy at the national level, and particularly by the framework for federal involvement in pollution control developed for the Clean Air Act of 1970. After the growth of the environmental movement in the 1960s and a greater awareness of the damage done to the environment, Congress began to feel greater pressure to address a range of environmental issues. First on the agenda was the Clean Air Act (CAA), passed in 1970. The CAA greatly expanded the federal role in air quality and established a regulatory framework that allowed the national government to impose air quality standards on states. The CAA also introduced the concept of primacy (see Crotty 1987), under which states could assume responsibility for the enforcement of federal regulations within the state. Primacy would be awarded to states who agreed to assume an enforcement role and were, in effect, granted authority by the national government to enforce federal regulations on behalf of the national government. The concept of primacy was meant to allow for the enforcement of a national standard for air quality while not making states irrelevant. However, if the states failed to enforce federal laws to the satisfaction of the federal government, the federal government could step in and assume authority for enforcement in the state, although revocations of authority are relatively rare (Shimshack 2014). This process significantly increased the role of the national government in environmental regulation, and effectively required all states to meet the requirements of national regulatory frameworks. While some states rejected primacy and refused to enforce national standards, the model of primacy became an important element of subsequent environmental legislation. Additionally, the Clean Air Act also set timetables for states to meet air quality goals, another "first" for environmental legislation.

Developing the Clean Water Act

Soon after the passage of the Clean Air Act, Congress began work on a new water quality program. Milazzo (2006) attributes the genesis of the 1972 legislation to a combination of environmental interest group pressure, the regulatory innovations contained in the CAA,[3] and the aspirations of Edmund Muskie (D-ME) to become president. However, Milazzo also notes that a combination of institutional legislative processes and competing agendas ultimately shaped the resulting legislation. For example, both the Senate Public Works Committee and the Senate Subcommittee on Air and Water Pollution invited a wide range of people to testify before their committee. In addition to the "usual suspects" of environmental group representatives and EPA officials, they also invited pollution control experts and industry representatives. However, also included were "...unlikely environmentalists, including men who designed ballistic missiles and an agency that built dams" (Milazzo 2006, 192). The point of this strategy was to create a culture in the committee of active participation from both senators and staffers, and to encourage open debate.

The Nixon administration proposed a bill, introduced in the Senate as S. 1014, that contained a number of interesting features. First, the proposal continued the use of water quality standards, but left the establishment of effluent standards to states. Federal enforcement would become a joint effort of the Department of Justice and the EPA. The proposal also authorized $2 billion per year for fiscal years 1972–1974, and a maximum of 55 percent federal cost share of projects. The bill also proposed a reallocation of grant funds: 45 percent on the basis of population; 10 percent for environmental need; and the remaining 45 percent for a variety of other uses (Lieber 1975, 32–33).[4] Finally, with the realization that state and local governments likely did not have the resources to apply to water quality on the same scale as the federal government, a companion bill (S. 1015) proposed the creation of an Environmental Financing Authority to be administered by the Department of the Treasury. The program would purchase municipal waste treatment bonds that could not otherwise be sold on the open market. The municipality would then guarantee payment on both principal and interest for the bonds. Ultimately the proposal was scrapped due to concerns about its reliance on ongoing appropriations (Lieber 1975, 33–34). However, the idea of a bond bank, albeit one administered by states, would later become an important element of the Water Quality Act of 1987.

By June 1971 the Senate committee developed the framework of a new bill. The proposal was to build on the existing framework, including the expansion of the regulatory framework passed the previous year in the Water Quality Act of 1970. To accompany this framework was an increase in funding for the construction grants program, although there remained disagreement about the magnitude of the increase. From the model provided by the CAA, Muskie also wanted to allow EPA to prohibit pollution discharges; to allow citizens to sue both polluters and the EPA; and to set deadlines to meet water quality standards. The White House and President Nixon were more circumspect; while they understood the pressures from the environmental lobby, they remained opposed to timelines and wanted to spend less than the $25 billion over five years proposed by Muskie. Nixon also wanted to reintegrate the US Army Corps of Engineers into the water quality regulatory regime,[5] although Muskie worried about administrative and regulatory overlap with the EPA (Milazzo 2006).

Joining Muskie on the Senate Public Works Committee was John Tunney (D-CA), at the time the youngest active senator when he won the seat in 1970. While Tunney was an inexperienced legislator, he contributed a concept to the conversation that became one of the centerpieces of the new legislation (Lieber 1975, 44). Building on the idea of a timetable in the form of the one found in the Clean Air Act, Tunney proposed an amendment that contained a clear target for a national water quality standard: That all waters in the United States "...achieve a level of purity sufficient to allow body contact with all the nation's waters, without health hazard, by 31 December 1980" (Milazzo 2006, 196–197). This idea, which would later morph into the "fishable and swimmable" goal in the legislation,[6] was unlike anything previously found in national water quality legislation. While in many ways a natural (and perhaps incremental) progression of the line of thinking about water quality policy, the notion of a specific timetable was fundamentally different. Milazzo (2006) suggests that, rather than being driven by a strong environmental ethic (or perhaps the strength of the environmental lobby), Tunney's proposal was instead a manifestation of systems thinking. Tunney had been in discussion with Simon Ramo, a well-known professor from the California Institute of Technology and entrepreneur known for his visionary thinking. Trained as an aerospace engineer, Ramo later began to apply systems thinking to social problems, and thence to environmental problems. A key element of this process is the delineation of clear goals; this is the element that led Tunney to think about how to articulate a goal to apply to water quality. Although Muskie was initially indifferent to Tunney's idea and concerned about the effects of the ambiguity of the concept as implementation guidance, it soon became clear that the political and environmental benefits of such an approach were significant (Milazzo 2006). After additional months of testimony, debate, and work, the committee settled on a "zero discharge" standard to describe the new national goal. The bill was finalized in committee and approved by the full Senate in early November.

The bill ultimately reported out by the Senate authorized $14 billion in new construction grants, another $2.4 billion to reimburse communities for past construction, and another roughly $1.6 billion for other uses (Davies and Davies 1975) – significantly more than the roughly $6 billion proposed by the Nixon administration.

Attention shifted to the House Public Works Committee. Its chairman, John Blatnik, was once again in a position to influence pollution policy. Although Blatnik's efforts had been opposed in the 1960s by members more interested in highways than wastewater, a series of electoral defeats left Blatnik the most senior member of the committee, and he ascended to the chairmanship (Milazzo 2006). However, not long after the House Public Works Committee began its work, Blatnik suffered a mild heart attack and was hospitalized. It then fell to Robert E. Jones (D-AL) to lead the committee process (Lieber 1975, 60). The Nixon administration had hoped to persuade the House committee that the Senate bill was unworkable and too ambitious, and worked to build a coalition of interests to rein in the Senate's excesses. Industry interests were fearful of federal standards, and were concerned that the Senate approach would make it difficult for industry to defend against lawsuits (Lieber 1975, 38).[7] While the Senate process had occurred largely behind closed doors, the House process was to be open and public. The administration's arguments were twofold: first, that to abandon ambient water quality standards was technologically infeasible; and second, that the cost of meeting the proposed standards would not be cost-effective – to meet the standards would require a portion of national resources far out of proportion to the benefits (Milazzo 2006, 227–228). The House bill produced by the committee in the winter of 1972 contained significantly weaker elements than its Senate counterpart: less reliance on technology-based standards, a longer timetable, and greater state discretion. However, the bill also authorized more money for the construction grants program ($18 billion, or $4 billion more than the Senate proposal), and also allowed up to a 75 percent federal share of construction costs (Lieber 1975). The bill that passed the full House in March 1972 was largely an unaltered form of the committee version.

Prospects for the conference committee seemed bleak, as the two versions were quite different (Lieber 1975). The areas of accord tended to revolve around the distributive elements of the bill, and Senate negotiators were willing to accept the House's larger authorization for the construction grants program, as well as the revamped grant allocation formula in the House version (Milazzo 2006). However, the points of greatest contention revolved around the imposition of water quality standards, and how to reach those standards. In addition, negotiators were at odds about how to incorporate (or separate) the National Environmental Policy Act's (NEPA) requirements for an environmental impact statement (EIS) as part of the wastewater permitting process. These were eventually resolved by including language to make clear that the EIS process could not allow licensing or permitting agencies to review or alter discharge limits (Milazzo 2006). After nearly five months of work, the conference committee reported its compromise bill to the House and the Senate. The bill was passed by wide margins in both chambers, but was vetoed by President Nixon (Adler, Landman, and Cameron 1993), who raised objections to both the estimated $24 billion cost of the program, and the standards for water quality contained in the bill.[8] The veto was quickly overridden the following day (Rinde 2017).

Lieber (1975, 82) suggests that Nixon's veto was, at least in part, an expression of his unhappiness with the Senate's rejection of H.R. 16810, "...a debt ceiling bill that would have authorized the president to restrict budget expenditures to $250 billion for fiscal year 1973." An element of Nixon's attempts to limit federal expenditures, Nixon was seeking greater executive authority to limit what he perceived as the spending excesses of the Congress. In essence, if Congress was not going to support his efforts to limit federal spending, he would not sign a new bill authorizing an expenditure in significant excess of his proposal for the program. This analysis is supported by Nixon's veto message, in which he challenged Congress as follows: "...a vote to sustain the veto is a vote against a tax increase. A vote to override the veto is a vote to increase the likelihood of higher taxes" (CRS 1973, 138).

Elements of the Clean Water Act

The bill that emerged from the conference committee was simultaneously an incremental step in a chain of water quality policies, and a significant departure from previous policy instruments. The bill contained five major sections, or titles: research and related programs; the construction grants program; standards and enforcement; permits and licenses; and general provisions (Stern and Mazze 1974). While a complete analysis of the bill is beyond the scope of this book, there are several major elements worth noting. First, the Act extended federal control over water policy to all navigable waters (Stern and Mazze 1974), a provision that solidified a long-building trend in water quality policy. Second, the bill gave authority to the US Coast Guard to determine the harmful effects of pollution from vessels and extended the Coast Guard's authority over the development of technology to control discharges from vessels (Stern and Mazze 1974).[9] These provisions were the result of several high-profile incidents, such as the oil tanker *Torrey Canyon* grounding off the southern coast of England in 1967 and a similar incident with the oil tanker *Ocean Eagle* off the coast of Puerto Rico in 1968, both of which had served to galvanize the environmental movement. The Act also extended federal enforcement action to cover all discharge of pollutants, and more clearly defined a violation as any discharge of pollutants. The Act also authorized several related areas for research, including the disposal of waste oil, the prevention and abatement of pollutants from sewers, pollution in the Great Lakes, pollution from mines, and pollution from economic growth centers. Missing from this list, but that would become an element of the 1987 legislation, is a focus on the Chesapeake Bay watershed, which by the early 1970s was considered significantly impaired (see Ernst 2010).

Although one of the major points of contention was the establishment of a water quality goal, the final bill established the goal of "eliminating the discharge of pollutants into navigable waters by 1985 and of attaining fishable and swimmable waters by 1983" (Poe 1995, 3). The orientation of the program was for point-source pollution, and the permitting and enforcement processes reflected this focus. The law also allowed states to assume full responsibility for meeting pollution limits and goals through the creation of State Pollution Discharge Elimination Systems, but if states failed to do so, they would be subject to the National Pollution Discharge Elimination System (NPDES). In other words, states could take control of their pollution programs if they desired, but if they did not desire, then they would be regulated by the federal government. The Act did not extend this authority to nonpoint sources of pollution, but Section 208 did provide for grants to states to control nonpoint pollution. A number of states designed nonpoint programs, but few were implemented due to a lack of funds (Poe 1995). The issue of nonpoint pollution would become a major point of contention in the next decade, but the 1972 legislation was the first instance in which nonpoint pollution issues became part of federal law.

The bill also greatly expanded the existing construction grants program, found in Title II of the Act. The legislation authorized a total of $24 billion over four years, a significant expansion of previous federal funding efforts. The provisions in Title II also changed the allocation formula for grants, and required states to develop priority lists for funding. The construction grants could now cover up to 75 percent of the total cost of the project, assuming the states financed at least 15 percent of the total cost (Lieber 1975). Title II also required states to use the best practicable technology available in the design of new construction, an element that ultimately proved difficult to meet and was subsequently the subject of amendment.

As noted above, Section 402 of Title IV established the National Pollution Discharge Elimination System. This section empowers the EPA, or through delegation, to states, to issue discharge permits to industrial and municipal polluters, based on the standards developed under Title III. In essence, Section 402 provided the compliance and enforcement mechanism to the Clean Water Act and serves as the tracking mechanism to track pollution discharges.

The NPDES thus tied the standards developed by states to an enforcement mechanism, and therefore to the Act's broader goal of "fishable and swimmable waters." Section 404 provided for wetlands protection, although the Act's overall impact on wetlands protection was negligible (Adler, Landman, and Cameron 1993). Nonetheless, the inclusion of wetlands as a water quality issue helped set the stage for a more comprehensive approach in future legislation.

Finally, another important element of the Clean Water Act is found in Title V, Section 505, which provides for citizen lawsuits against polluters and, if no action is taken, against enforcement agencies as well. This provision adds another method of redress for citizens and means that lawsuits can be brought by entities other than the Department of Justice. Moreover, citizens can now seek court authority to force implementing agencies to take action where agencies are reticent to act. Although there have been some notable cases brought under the auspices of this provision (see Andreen 2004; Flatt 1997; Pedersen 1988), Section 505 would become much more important (and controversial) in the 1990s and 2000s as citizen groups used Section 505 authority to sue the EPA and state environmental agencies to enforce nonpoint pollution standards, particularly regarding the creation and enforcement of Total Maximum Daily Load (TMDL) standards (see Houck 2002).

Summary

The Clean Water Act of 1972 is rightly considered a turning point in the history of water quality legislation in the United States. Although it was the natural outcome of decades of previous policy development, the 1972 legislation brought to fruition a clear goal for water quality policy, along with the resources necessary to address the problem. The elements of the CWA are also entirely consistent with the broader policy streams in place at the time. The solution stream added the elements of specific goals to the legislation, as well as an increased role for the federal government at the expense of state discretion. The increased focus on environmental concerns in the politics stream also allowed for a more proactive approach to environmental policy, one that took a much more aggressive approach to environmental protection and pollution cleanup than would have been possible even five years previously. The tremendous growth in both the size and influence of the environmental lobby provided a clear foil for more traditional industrial and municipal interests, and provided a living example of David Truman's (1951) interest group politics. The presence of an effective policy entrepreneur, in the form of Edmund Muskie, also helped push the process along. The net result was the most sweeping change in water quality policy to date. Yet, for all its change, the underlying elements of the Act rested heavily on previous policies. Perhaps the distinguishing factor is that, while previous policies tended to extend one (or perhaps two) changes each time, the CWA72 extended the reach and importance of several elements all at once. Coupled with the addition of specific goals for water quality, the 1972 legislation stands as the most important water quality policy statement of the century and would, in turn, inform and guide the changes that followed over the course of the next fifteen years.

Redefining the Streams: 1977–1986

While the stated intent of the 1972 amendments was to *assist* municipalities in meeting their water treatment needs, it became clear by the late 1970s that municipalities were using the grants program as a substitute for state and local funding (see EPA 1984; Heilman and Johnson 1992, 36). Indeed, by 1976 state and local spending for wastewater treatment works (WTWs) had reached its lowest point in over two decades (EPA 1984). This state of affairs continued into the early 1980s, although the total amount spent on WTW construction began to decline, due in part to the effects of the 1977 amendments. The net result was that the backlog of needs was not significantly reduced. Although many facilities were built or upgraded during this period,[10] significant wastewater treatment needs remained unmet.

The Clean Water Act of 1972 was first amended in 1977. The amendments were relatively minor, although there were two important changes brought about in this round of changes. First, the 1977 legislation more carefully and fully defined the type of pollutants that were covered under the Act, particularly in terms of toxic substances (Poe 1995). This action brought the CWA in line with the recently enacted Toxic Substances Control Act of 1976. Second, and much to the relief of states and localities, the amendments postponed the original deadlines for achieving the "fishable and swimmable" standards set in the original legislation. This was driven by a growing consensus in both EPA and Congress that the development of technology was insufficient to make the originally proposed deadlines viable, a repudiation of the 1972 requirement for the use of "best practicable technology" under Title II. The 1977 legislation also extended CWA authorizations through 1981. The CWA was amended again in 1981, again with three basic changes. First, the Act extended authorizations through fiscal year 1986. Second, the deadlines for meeting the "fishable and swimmable" requirements were again postponed, but not removed from the legislation.[11] Third, the Act made some minor changes to the construction grants program (Poe 1995), mostly in terms of cost share requirements. Grants for project planning and design were disallowed, and the number of eligible categories was reduced from previous legislation. The net result was to focus more clearly on current needs, rather than on future needs (EPA 1984, 1-2).

The Election of Ronald Reagan and Changes in the Streams

The election of Ronald Reagan to the presidency in 1980 brought a significant change of focus and redefinition about the role of government. Reagan was elected on the basis of a message of smaller government, a broader role for the private sector, and a renewed interest in states' rights (Johnson and Heilman 1987; Palmer and Sawhill 1984). Known collectively as the "New Federalism" or Reagan federalism (Conlan 1988, 3), the underlying issue was public resources and increasing demands for public services. In an effort to stimulate economic growth in the early 1980s, the Reagan administration effected a radical change in the federal tax structure (Birnbaum and Murray 1987; see especially chapter 3). The dual purposes of these revisions were (1) to reduce the tax burden on corporations and individuals, and therefore stimulate investment and (2) provide incentives to attract private participation in public works projects.

There were three main elements to Reagan's "New Federalism."[12] First, the Reagan administration pushed for legislation that reduced the tax burden on both individuals and businesses. The Economic Recovery Tax Act (ERTA) of 1981 and the Tax Equity and Fiscal Responsibility Act of 1982 were two early versions of this element. The ERTA is particularly important in this respect, in that it provided substantial tax incentives for private companies to invest in capital-intensive public infrastructure projects. While a number of these incentives were ultimately repealed in later legislation (see Heilman and Johnson 1992), they formed the basis for greater private sector involvement in public infrastructure needs. In addition, by mobilizing private sector capital, Reagan hoped to reduce the federal deficit (and the federal involvement in infrastructure provision).

Concomitant with this first element, Reagan sought to redefine the role of the national government in the United States. Dismayed by a national government that he believed had become too big, too powerful, too intrusive, and too expensive, Reagan attempted to make government smaller and less intrusive. He attempted to reduce federal regulatory regimes and to promote privatization for the provision, production, and delivery of many goods and services previously offered by the national government. This strong philosophical belief in the fundamental superiority of capitalism as an allocation mechanism for goods and services proved a powerful force in Reagan's politics.

Third, Reagan sought to devolve both authority and responsibility for many federal programs to the states. A strong proponent of states' rights, Reagan believed that states were better suited to make policy choices than was the national government. To this end, Reagan championed the use of block grants (rather than categorical grants), or simply to devolve programmatic and fiscal responsibility to the states. As pressures to reduce the federal deficit grew (see Thelwell 1990), devolution often meant a cessation in federal fiscal support but retaining requirements for states to meet national policy standards and goals (Conlan 1986).

Taken together, these policy streams signal a shift in metapolicy that had far-reaching effects in many policy sectors. In terms of water policy, these changes in the problem, solution, and political streams are reflected directly in the Water Quality Act. The streams no longer contained the elements of a federally led, federally funded, open-ended program to support water quality. As early as the late 1970s, a growing consensus in Washington, DC suggested that the construction grants program was not producing the intended results, a discussion that led, in part, to the changes in the construction grants program in the 1981 amendments. Earlier studies had indicated that the construction grants program was not, in itself, sufficient to meet the demand for water quality resources. However, it was also the case that there seemed to be little appetite (or political opportunity) to roll back clean water regulations. A focus on unfunded mandates would not manifest itself for a decade or more, so the policy question at hand was how best to fund wastewater needs in the states at a level to meet state needs.

The 1984 EPA Task Force Report

The 1981 amendments reauthorized the construction grants program through fiscal year 1986. Realizing that Congress and the White House would want to consider options to the construction grants program as part of a reauthorization process, EPA administrator William Ruckelshaus convened a task force to examine the federal role in water quality, and specifically the federal government's role in municipal wastewater treatment, in October 1983. The task force was led by Jack Ravan, the assistant administrator for water, and consisted of nine other senior EPA employees. Three of the EPA members were from EPA regional offices, representing Regions IV, V, and VI.[13] The final member was Reginald LaRosa, the president of the Association of State and Interstate Water Pollution Control Administrators (ASIWPCA),[14] who served in an *ex officio* capacity on the task force. The task force also worked closely with the Management Advisory Group (MAG) to the Construction Grants Program, a sixteen-member group that brought together EPA staff, industry experts, engineering experts, financial experts, local government officials, and representatives of environmental and citizen groups from across the nation (EPA 1984, 1-3-4). The charge to the task force was to develop alternatives to the current construction grants program that would alter the federal role in water quality. In February 1984 EPA published a "call for papers" in the Federal Register that invited interested parties to submit input on a range of funding options and evaluative criteria (EPA 1984, 1-3).[15] The MAG provided its recommendations to the task force in May 1984, and the task force released its findings in December 1984. A review of the task force's report illustrates the effects of Kingdon's policy streams on the conversation: the assumptions made by the task force; the kinds of solutions offered; and the ways in which the problems were defined.

Problem Definition

The task force's report placed a great deal of emphasis on problem definition. First, the report dedicated an entire chapter (Chapter 3) to a discussion of the federal funding effort in comparison to state efforts. In short, the task force noted that federal spending on wastewater treatment had increased dramatically since 1972 as a result of the Clean Water Act. At the same time, state and local government funding for wastewater treatment declined sharply beginning in 1972; by FY 1974, the national government was spending more on wastewater treatment than were states. This stood in marked contrast to the pattern prior to 1972, when the bulk of the resources spent on water quality came from state and local sources. In FY 1976, federal funding totaled $5.38 billion, while state and local funding nationwide failed to reach the $1 billion mark. This constituted a reduction in state and local effort of more than 75 percent (in constant 1982 dollars) (EPA 1984, 3-1). In effect, the CWA had created an incentive for state and local governments to replace their funds with federal funds, something the federal government was clearly willing to do.

A second element of problem definition involved a growth in the need for water quality funding. The report noted an 11 percent growth in the US population between 1972 and 1982, with more growth to come (EPA 1984, 3-3). Indeed, the report cited EPA's 1982 Needs Survey, which reported a backlog of projects totaling nearly $93 billion; this need was expected to increase to over $119 billion by 2000 (EPA 1984, 3-8). Current core needs[16] from the 1982 Needs Survey alone totaled over $35 billion (EPA 1984, 3-11). To meet this need at current federal funding levels would require another thirteen years of federal funding, not counting the effects of inflation, current ongoing projects, or state administrative costs (EPA 1984, 3-8). The task force's report makes clear this level of funding is probably unrealistic, prompting the need for alternative mechanisms.

Though not explicitly mentioned as a "problem" in the report, the task force also incorporated a desire for greater private sector involvement in the water quality program. The Executive Summary at the beginning of the report suggests that privatization "…could become an important factor in financing wastewater treatment facilities for population and industrial growth" (EPA 1984, xi). At the time the task force was working, the tax incentives enacted early in the Reagan administration were still in force, and the hope was to attract more private investment into public infrastructure through the implementation of these incentives. The one report originating outside EPA included as an attachment in the report was provided by Peat, Marwick, Mitchell, & Co. (now KPMG), a private accounting firm. The analysis offered five different financing schemes; two of the schemes assumed a private services contract model (EPA 1984, 2 [Attachment 3]). The task force dedicated some four pages of the report touting the potential benefits of privatization, largely by contrasting the benefits to the costs of continued governmental funding – in short, defining the solution in terms of a problem. Additionally, the in-depth discussion of leveraging in the report (discussed later in this volume) relied heavily on a private sector model of finance.

Areas of Consensus of the Task Force

The report touted an "astonishing" (EPA 1984, ix) degree of consensus, both in terms of the underlying issues surrounding the federal role in water quality, as well as the proposed solution for future legislative initiatives. As listed in the Executive Summary, these areas are as follows (EPA 1984, ix):

(1) A gradual transition from Federal responsibility to State and local self-sufficiency should be pursued.
(2) While Federal aid continues, its level should be certain and States should be able to deploy funds flexibly.
(3) Continuing and future wastewater treatment needs must be made an element of any new funding scenario.
(4) Delegation of Construction Grants Program responsibilities to the States has been successful and should continue.
(5) Major changes in standards and compliance deadlines efforts are counterproductive.
(6) Federal funding strategies must promote compliance and be supplemented by strong enforcement actions.
(7) Funds should be distributed equitably to meet core treatment needs first [emphasis in original], and States should be given the flexibility to address project affordability issues.
(8) The funding mechanism chosen must provide for both short-term and long-term financial "leveraging" of available funds. That is, the earning power of each funding dollar must be maximized.

Taken together, these eight areas of consensus represent a remarkable fidelity to the changes in the policy streams in the 1980s noted earlier in this chapter. The themes of a smaller role for the national government, increased responsibility for the states, program flexibility, and affordability are all consistent with the metapolicy (Dror 1968) of the Reagan administration. At the same time, the reaffirmation of the "command and control" regulatory framework already in place is equally remarkable. National water quality still matters, and EPA is still invested in the existing regulatory regime.

Developing Options

Chapter 5 of the task force's report offered a total of five different policy options for consideration. Each of the five options was accompanied by a list of advantages and disadvantages. The chapter also lists the criteria against which these options were reviewed; the list includes effectiveness, efficiency, equity, and feasibility.[17] The options were also assessed for their potential negative impacts, including the creation of new federal off-budget entities to manage a new program; the creation of significant new financial liabilities for the federal government; the possibility for significant increases in federal tax losses; and the possibility that an option would further incentivize state and local governments to reduce their funding share (EPA 1984, 5-1-2). The first option offered was to rely on financing through municipal bonds, with or without privatization. While this option would increase state and local discretion and control, the financial costs (mostly due to high bond interest rates, and political restrictions on general obligation bonds) were deemed too high to make this option feasible (EPA 1984, 5-3-4).

The second option was to rely on municipal bonds with credit enhancements. This plan would rely on the issuance of municipal bonds to be guaranteed by the federal government. While the plan was envisioned to reduce interest rates paid on those bonds, especially for financially strapped communities, there were fears that communities could intentionally default on the bonds, essentially receiving "free" financing (EPA 1984, 5-5). The task force concluded that this option would do little to promote self-sufficiency for state and local governments. The third option offered was a federal loan program. In this model, the federal government would loan money to communities at a subsidized interest rate. Loans would only be available for projects currently available for federal assistance. While this plan would have the least negative effect on federal tax losses, there were concerns that the costs to the federal government could exceed the costs of the construction grants program. As with the credit enhancements option, this option would also do little to promote self-sufficiency, and would also limit flexibility for state and local governments (EPA 1984, 5-8-9). The fourth option was a federal grant program; in effect, a continuation of the construction grants program. The advantages of the program were that the program was already well-established, effectively managed and delegated; the funds went specifically to wastewater needs; and state control over priority lists for funding provided states some level of control. The disadvantages largely centered around a lack of flexibility and funding alternatives absent the federal program. However, the task force also pointed out that a federal grant program would do nothing to enhance state financial capacity:

The grant program does not promote long-term capital formation for State and local self-sufficiency. Since Federal grants substitute or replace State/local expenditures, the program does not provide adequate incentives to transfer the true cost of treatment and collection to the users of the system and does not encourage the development of State/local sources of funding.

(EPA 1984, 5-10)

This point illustrates the importance of one of the key elements of the push for a revolving loan fund program. One of the important goals was to move the cost of wastewater treatment off the federal budget and to pass the costs to states and municipalities, while simultaneously maintaining the federal regulatory regime.

The final option, and the centerpiece of the task force's report, was for the federal government to provide capitalization grants for the establishment of state-managed revolving loan fund programs. The program would combine a federal capitalization grant with state matching funds, to be placed in an account managed by the state. The state would make loans to communities from this fund, which would be repaid over a twenty- to thirty-year period. Because the loans would carry an interest rate, the value of the state fund would grow over time; the proceeds from previous loans would be used to make new loans. In this manner, the fund would grow over time, and eventually would make each state self-sufficient in terms of meeting its wastewater needs. The focus of the task force's argument was on the self-sufficiency of states, and the ability to build state capacity. In addition, because states would be making funding decisions, states would have greater flexibility to determine and meet their own needs without federal government intervention or guidance.

In spite of their obvious preference for a revolving fund program, the task force did identify several potential disadvantages to such a program. First, unless states supplemented the revolving funds with additional monies, the funds initially would be too small to provide the same level of subsidy to communities. Second, the need for financial solvency in the programs would likely mean that at least the initial round of loans made would be dependent on the financial capacity of the recipient communities, with a clear emphasis on the ability to repay the loan. This would likely leave communities in financial difficulty on the sidelines, and unable to meet their water quality needs. Third, the funds would require adequate capitalization to be effective. This would necessitate both adequate initial capitalization from the federal government, as well as the ability of states to provide additional funds. Finally, the task force acknowledged that the nature of a revolving loan fund would require significant financial expertise on the part of states to be administered efficiently and effectively (EPA 1984, 5-15).

The report also recommended the creation of a transitional program in the event that policy makers chose an option other than the status quo construction grants program. Given that the option of choice for the task force was the creation of a revolving loan fund, the task force recommended a period of several years in which the construction grants program would continue, but during which capitalization funds would be provided to create the state-administered revolving loan programs. The task force recommended that capitalization grants would be offered from 1985 to 1994, after which states would assume full responsibility for the program. The task force specifically pointed out that the capitalization grant period needed a clear cutoff point, so that states would be encouraged and incentivized to assume responsibility and provide funding continuity and not believe that federal funding would continue past that date.

The task force laid out a transitional program in some detail. For example, the task force recommended a specific pattern of funding authorizations, beginning at $2.4 billion for the first six years of funding (FY 1985–FY 1991) and dropping to $600 million in the final year (FY 1994).[18] The report also recommended that the allotment formula to distribute program funds to states be the same as currently in force, and that the same formula continue through the life of the authorization period. This would give states the ability to plan based on a known amount of funding. The transitional program also recommended guidance for the use of set-asides for program administration, the timing of state funding decisions, and the requirement of a 20 percent state match for each year of federal capitalization funding. Finally, the task force recommended specific elements for state assurances to the federal government and to local governments, including allowable uses for the funds, financing mechanisms, managing for environmental results, and compliance with applicable federal law (EPA 1984, 6-3-6-11).

The other noteworthy element of the task force's proposal was the acknowledgment that the initial federal capitalization was unlikely to be sufficient for states to meet current and future needs. To this end, the task force recommended that states should be allowed to adopt leveraging strategies to increase the amount of funds available for loans. States could make the decision whether to operate a leveraged or unleveraged fund, but the federal law should allow states to use their capitalization grants as reserve funds to support a leveraging strategy (EPA 1984, 6-12). The task force suggested that a leveraging strategy could increase available funds in the state for loans by as much as a factor of five (EPA 1984, 6-12). The issue of leveraging became a centerpiece of program implementation, and a point of some contention between the competing program goals of the long-term viability of the state programs and the ability to meet current and future needs.

The final issue addressed by the task force's report concerned the status of the various set-asides incorporated into the 1972 Clean Water Act. An analysis of the effects of the set-asides written by EPA's Office of the Comptroller, included in the report as an attachment, served as the basis of the task force's recommendations regarding the set-asides. These set-asides included up to 4 percent of a state's construction grants allotment for state management assistance; a 1 percent set-aside for state water quality management planning; a 4 percent set-aside for rural communities;[19] and a set-aside ranging from a minimum of 4 percent to a maximum of 7.5 percent to be used to promote alternative or innovative technologies (EPA 1984, 3-4 [Attachment A]). In this case, rather than make specific recommendations, the comptroller's report contained seven different options, ranging from the abolishment of set-asides, to making them optional for state participation, to the creation of new set-asides, to providing states more flexibility under the current programs. Each of the options was accompanied by a discussion of advantages and disadvantages, all based on the task force's consensus goals. As we will see, Congress considered these options when developing the successor to the Clean Water Act, but the final legislation treated the former set-asides differently than had been the case in the 1972 legislation.

Summary

The 1984 task force report thus became the basis for the Water Quality Act of 1987. The ways in which the issues were defined, as well as program goals, structure, and requirements were drawn almost verbatim from the 1984 report. The ways in which the problems and solutions in the report were framed were themselves driven very much by the prevailing politics, problem, and solution streams as they existed at that moment in time. It is likely that the high degree of consensus among the task force recipients also influenced congressional policy makers and made it easier for them to reach consensus around a new approach to water quality. Indeed, as we will see, the resulting legislation enjoyed a high degree of bipartisan support in Congress, no doubt due, at least in part, to the congruence of the proposal with the prevailing policy norms in Washington. Although the nature of the program represents something of a larger change from the funding approaches of the previous four decades, the changes in the politics stream made a larger change palatable.[20] Armed with this report, Congress began work in 1985 to further amend the Clean Water Act.

A New Approach: The Water Quality Act of 1987

One of the first items of business in the new 99th Congress was H.R. 8, the Water Quality Renewal Act of 1985, introduced by James Howard (D-NJ). The initial bill contained a number of changes from the Clean Water Act and its subsequent amendments, but chief among the changes was the inclusion of a proposal for a revolving loan fund to replace the construction grants program. The bill was referred to the House Committee on Public Works, and thence to the Subcommittee on Water Resources. The subcommittee began hearings in late April and held a markup session in late May. The bill was again marked up, amended in the full committee, and reported back to the full House in early July. The House Rules Committee reported its rule to the House on July 15, and the rule was accepted by the House on July 18. The bill was passed in the House by a bipartisan vote of 340-83 on July 23. Meanwhile, a similar bill, S. 1128, was introduced in the Senate in mid-May 1985 by Senator John Chafee (R-RI). The bill was generally similar in content to H.R. 8, and was referred to the Senate Committee on Environment and Public Works. The bill spent relatively little time in committee and was passed by the full Senate on June 13 by a vote of 94-0.

However, significant differences between the two bills, particularly over authorization amounts, led to a protracted conference committee process. The Senate version included $190 million for a combination of program administration and funds to study pollution in freshwater lakes, while the House bill authorized $140 million for the same two functions. However, the House bill also authorized an additional $150 million for the development of areawide waste treatment management ($50 million), a rural clean water program ($50 million), and to implement interagency agreements related to clean water ($50 million). The Senate version of the bill authorized $9.6 billion to continue the construction grants program through FY 1990 as a bridge between the construction grants program and the new revolving loan fund program. The House bill authorized the construction grants program at then-current funding levels; $2.4 billion per year from FY 1986 to FY 1990 (a total of $12 billion). The House bill also provided increased shares to the Virgin Islands, American Samoa, the Northern Mariana Islands, and the Trust Territories to better meet the water quality needs of these areas. The largest differences in authorizations were related to Title VI, the new state revolving fund program. While the Senate version authorized $8.4 billion from FY 1989 to FY 1994, the House bill authorized $9 billion between FY 1986 and FY 1994. The conference bill ultimately included the Senate authorization, and kept the shares to the territories and protectorates at current levels.

The versions also differed in terms of special projects. Each version had several demonstration or special projects not included in the other; the final bill included nearly all the projects from both bills. The Senate bill did not include language regarding the total federal share allowed under the "bridge" construction grants program, but the House version, ultimately adopted in the conference report, allowed a continued 75 percent federal share for projects that received grants prior to October 1, 1984. The Senate bill offered a revised allotment formula for construction grants for fiscal years 1989 and 1990, while the House version retained the existing allotment formula. The conference bill established a new allotment formula for fiscal years 1987–1990 but retained the current formula for FY 1986.

The Senate version of the bill authorized $300 million over three years for states to implement nonpoint-source projects, with a federal share of up to 75 percent of project costs. The House version authorized $150 million over five years, with a cap on the federal share of 50 percent of project costs. The conference substitute authorized $400 million over four years, with a limit of 60 percent of project costs. A final difference worthy of mention was a House provision that would amend Title V to exempt agricultural stormwater runoff from the definition of "point-source" pollution.[21] The Senate bill had no comparable language, and the conference bill adopted the House language.

In terms of the structure of the new state revolving fund program, the House and Senate versions of the bill were largely the same in terms of structure, although the language was different in each. The Senate version tended to focus more on accountability: state Intended Use Plans (IUPs), accounting controls, and annual state reports were required in the Senate version. The Senate version also called for a 15 percent state match to the federal funds received, and binding commitments must be entered into within twelve months of the federal grant. The House version required a 20 percent state match and broadened the allowable uses for CWSRF funds as compared to the Senate version. The conference bill effectively combined language from both versions, although it adopted the House match requirement of 20 percent. Both versions also contained a "first use" clause, which required states to use all funds first to assure progress toward compliance with enforceable deadlines, which placed the clear emphasis of the program on environmental need.[22]

The conferees agreed on a final version of the bill on October 10, 1986, and the conference bill was filed in the House five days later. The House voted to approve the conference report by a vote of 408-0, and the Senate followed suit the same day by approving the conference report by a vote of 96-0. The final bill was presented to the president on October 25. Congress then adjourned for the year as members returned to their districts in anticipation of the upcoming midterm election. However, President Reagan, having publicly objected to the cost of the new bill, chose to exercise a pocket veto on November 6.[23] Since Congress had adjourned for the midterm elections, the bill was effectively dead. Given the overwhelming support in both chambers for the bill, it was likely the president would not have the last word.

Indeed, on the very first legislative day of the new Congress H.R. 1, the Water Quality Act of 1987, was introduced in the House by James Howard (D-NJ), the sponsor of the previous bill. However, this time he was joined by 167 cosponsors in the House. The bill was identical to the legislation vetoed by the president some two months earlier. In this case, the bill was approved by the House on January 8, 1987, by a vote of 406-8. The Senate approved the bill on January 21 by a vote of

93-6 and the legislation was presented to the president the same day. Nine days later President Reagan again exercised his veto power, again citing cost as the issue.[24] Congress was not deterred; the House voted to override Reagan's veto on February 3 by a vote of 401-26, and the next day the Senate overrode the veto 86-14.[25] The Water Quality Act of 1987 was now law.

In spite of pushback by fiscal conservatives, the Water Quality Act received overwhelming support in both chambers of Congress. The greatest expression of resistance was seen in the House at the point at which the 1985 version was passed; those opposed were a collection of fiscal conservatives, those who opposed the loss of the benefits of categorical grants, and some members influenced by manufacturing interests. At the end of the day, however, even that resistance crumbled as the 99th Congress accepted the first conference report with unanimous votes in both chambers. Water quality still mattered as a policy issue, and the new program fit the current politics stream in that it reduced (and, in theory, ended) the federal funding commitment for water quality, gave greater autonomy and control to states, and reduced the federal role in water quality. The significant elements of the 1984 EPA Task Force's report had materialized in the form of law, and the same forces that shaped the 1984 report were reflected strongly in the discussion and debate leading to the passage of the Act.

The change in focus of US water quality policy was again incremental, yet the abandonment of the construction grants program in favor of a state loan fund represented a significant change in federal-state relations in water quality. The press for states' rights had reached full voice, and states would now shoulder the responsibility for meeting national water quality standards without the prospect of continued federal funding. The phased withdrawal of the construction grants program seemed to soften the initial blow to a point, but some observers were already beginning to question whether the authorizations would prove sufficient to capitalize state funds (see Holcombe 1992; Travis, Morris, and Morris 2004). Still the victory for Reagan federalism was significant, and the Water Quality Act of 1987 remains one of the clearest expressions of the theory behind Reagan federalism. The politics, problem, and solution streams collided at a point at which all the elements of Reagan federalism were in play. However, the one element of Reagan federalism that failed to materialize in the WQA87 was a reduction in the federal regulatory role in water quality. Deregulation was an important element of Reagan federalism, and the WQA left the previous regulatory structure intact. Moreover, it expanded the federal role by placing new emphasis on nonpoint pollution. Although Reagan noted his dissatisfaction with the expansion into nonpoint-source pollution in his second veto message, little mention was made of the existing regulatory framework. Still, the WQA embodies a great many of the elements of Reagan federalism.

Post-Water Quality Act Legislation: 1988 to the Present

Although the original concept for the Water Quality Act envisioned federal capitalization grants to end by FY 1994, it became increasingly clear that the resulting funds would not be adequate to meet water quality needs in most states. Under pressure from the states, Congress simply appropriated additional funds each year. The amounts varied, rising significantly during 2008–2010 as a result of the stimulus package enacted to help stabilize the economy after the recession began in late 2007 (see Copeland 2014).

The changes brought about post-1987 were highly incremental in nature, perhaps more incremental than the period from 1948 to 1956. The past twenty-five years have been marked by a high degree of partisanship in Congress, during which a number of policies have languished in legislative limbo. In many ways, this period of legislative history is explained by Lindblom's (1959) increment-alism: When partisanship is high, the increments of change get smaller, since it is easier to reach agreement on small changes. What is somewhat more remarkable is that appropriations for the CWSRF have continued past FY 1994, even absent congressional authorization (see Copeland 2014, 1–2). A divided Congress' unwillingness, or inability, to reauthorize the 1987 legislation, even in the face of high demand from states and communities, is indicative of yet another change in the political stream. In spite of these changes, the 1987 legislation has undergone several changes, albeit small ones.

A minor change was made to the revolving loan fund program with the passage of the Safe Drinking Water Act (SDWA) of 1996. With the revolving loan fund model of environmental infrastructure finance still in mind, Congress chose to adopt the revolving loan fund model for drinking water infrastructure projects as well. More importantly, the SDWA allowed states to comingle the Drinking Water State Revolving Fund (DWSRF) monies with those from the now renamed Clean Water State Revolving Fund (CWSRF) program. The specifics of this provision did not alter the clean water portion of the program or provide significant additional funds for clean water; the most important change was that it allowed states to use limited amounts of funds from one program in the other. As a result, a number of states also moved implementation of both programs, at least in terms of the financial elements, into the same agency.

Another round of minor program changes occurred in 2014 as part of the Water Resources Reform and Development Act (WRRDA; P.L. 113–121). Although the bulk of the 2014 legislation was focused on the US Army Corps of Engineers and amended previous versions of the Water Resources Development Act, the WRRDA also contained several provisions relevant to the Water Quality Act. First, the WRRDA required states to ensure that the CWSRF funds will be maintained such that they are available in perpetuity. It also required states to redirect any fees or charges to loan recipients to administration of the state program. As was the case with other legislation passed in this period, the amendments also prohibited the funding of any project unless the iron and steel used in the project was produced in the United States.[26] The amendments broadened the definition of eligible projects to include projects that were energy- or water-efficient, and expanded the ability of states to offer loan repayment terms up to a period of thirty years. Finally, the Act authorized a pilot program, the Water Infrastructure Finance and Innovation Act (WIFIA), that would provide federal loans and loan guarantees for large regional water projects including drinking water and wastewater (Copeland 2014, 6).

Congress again made negligible changes to the program structure in 2018–2019. In P.L. 115–436, the Water Infrastructure Improvement Act, Congress directed the EPA to create an Office of the Municipal Ombudsman to provide technical assistance to municipalities. This new office was located in the Office of the Administrator rather than in the Office of Water, the traditional source of technical assistance for water issues. The Act also required the EPA administrator to promote the use of "green infrastructure" in the planning, research, technical assistance, and funding guidance of the Clean Water Act provisions. These changes were approved by both chambers in December 2018 and sent to the president in January 2019; the president signed the legislation on January 14th.[27]

Although authorizations ended in FY 1994, the underlying policy instrument created under the Water Quality Act of 1987 has changed very little in the intervening three decades since its passage, marking the longest period since 1948 that Congress had not significantly updated or altered water quality policy. The changes since 1987 have mostly focused on the types of projects eligible for CWSRF assistance, and more technical issues related to program implementation. But the framework laid out in the 1987 legislation has remained remarkably intact, and serves as a prime example of Reagan federalism. Although the original legislation failed to meet Reagan's (far-reaching) vision for the federal–state relationship in water quality, the WQA nonetheless embodies the underlying principles of Reagan's vision, and represents a significant redirection from previous policy thinking.

Summary

The history of water quality policy in the United States is one of incremental change driven by the larger political forces in play. The federal involvement in this policy arena began in a very small way, and for the next two decades or so the basic framework was expanded and strengthened. The Clean Water Act of 1972 was notable for both its increase in the national commitment to water quality, as well as the imposition of specific national water quality goals. Although these goals proved unworkable, the Clean Water Act nonetheless solidified the viability of national standards for clean water. The 1987 legislation was somewhat less far-reaching in its scope, but the shift in responsibility for funding from the national government to state governments represents the first major conceptual change in water quality funding since the imposition of federal water quality funding in 1948. As was the case in the early period of wastewater policy, the period from 1972 to the present is marked by a largely incremental approach to water quality policy (see Table 4.1). The second half of the twentieth century is particularly

Table 4.1 *Major elements of later water quality legislation, 1972 to date*

Act	Year	Major Elements
Federal Water Pollution Control Act Amendments[*]	1972	Greatly expanded size of the construction grants program; set goal of "fishable and swimmable waters by 1983"; created NPDES permitting mechanism; extended federal control over water policy to all navigable waters; developed watershed-based approach to water quality; gave US Coast Guard authority to determine effects of pollution from vessels; allows for citizen lawsuits against polluters
Clean Water Act	1977	More carefully defined classes of pollutants covered under the CWA; postponed the deadline for achieving the "fishable and swimmable" goal; extended Title II authorizations through FY 1981
Clean Water Act Amendments	1981	Extended appropriations through FY 1986; further postponed "fishable and swimmable" goal; cost share changes for Title II grants
Water Quality Act	1987	Creation of Title VI (CWSRF); elimination of Title II authorization after FY 1990; changes in funding formula; new emphasis on nonpoint-source pollution; creation of Chesapeake Bay and Great Lakes program offices in EPA; permit changes for discharge into marine waters; new compliance dates for meeting BAT requirements; creation of Small Flows Clearinghouse; expand criminal and civil penalties for polluters
Safe Drinking Water Act	1996	Creation of DWSRF, and ability to comingle CWSRF and DWSRF funds
Water Resources Reform and Development Act	2014	Requires states to maintain CWSRF funds in perpetuity; requirement to use US-produced iron and steel in CWSRF-funded projects; broadens eligibility CWSRF funds to include energy- or water-efficient projects; provides loan guarantees for regional water projects
Water Infrastructure Improvement Act	2019	Creation of Office of the Municipal Ombudsman in EPA; promote the use of "green infrastructure"

*Although formally titled as noted here, the 1972 legislation is widely known as the Clean Water Act.

notable, however, for incrementally larger changes brought about by the Clean Water Act in 1972 and the Water Quality Act in 1987. While certainly not revolutionary in policy terms, each of these two Acts served to change the nature of water quality policy.

As with any policy, there were important assumptions underlying the 1987 legislation, all of which are ripe for examination. Do states have the administrative and technical expertise to design, implement, and administer a complex revolving loan fund? Are states willing to invest more state funds into water quality projects? Would communities be willing to choose loans from states in lieu of borrowing from the open bond market, or self-funding? Could the new program serve the needs of the kinds of communities Congress wished the program to serve, or would the funds be pushed to communities best positioned to afford the loans? Would states truly become self-sufficient in water quality funding? With more than thirty years of empirical evidence at hand, we are able to address these assumptions and the questions they raise and determine the long-term viability of the revolving loan fund model.

5

Features of the Water Quality Act of 1987

The Water Quality Act (WQA) of 1987 was, in many ways, an incremental step forward from the Clean Water Act (CWA) of 1972. Much of the language of the WQA was retained in the same form as found in the CWA, and many of the changes represent minor changes to existing programs, new dates to meet certain program goals, and similar amending language. The overall purpose of the Acts is the same, as are many of the programs contained within the two Acts. Indeed, outside of the inclusion of a new program to fund wastewater treatment facilities in states, the two Acts are very much alike. As we will see, even with the addition of a new funding mechanism in the form of a revolving loan fund program, the existing funding program of categorical grants remained in the WQA. Moreover, not only did the previous grant program remain in the legislation (and authorized for three additional years), Congress continued to appropriate funds under the categorical grant program for more than two decades after the authorization had expired.

Perhaps the biggest, and arguably most important, change is the addition of a new title to the Water Quality Act. Title VI creates the Clean Water State Revolving Loan Fund (CWSRF) program, which fundamentally changes the relationships between the national government and states (and, not coincidentally, local governments as well) in the arena of national water quality policy. The original funding mechanism of the CWA, Title II, described a categorical grants program that allowed Congress, through the Environmental Protection Agency (EPA), to make grants directly to eligible municipalities, effectively bypassing state decision makers in the process. This was a very deliberate decision made in the CWA, and had been modeled on a program first enshrined into law as a part of the Federal Water Pollution Control Act (FWPCA) of 1948 and funded in the 1956 amendments to the FWPCA. Between 1948 and 1987 the construction grants program had served as the primary funding stream provided by the national government to construct wastewater treatment facilities. The advent of the CWSRF program, as detailed in Title VI, switched federal efforts from a categorical grant program to a block grant program.[1] The ramifications of that switch are the focus of this book.

Although the "lead story" of the WQA is the inclusion of Title VI, to understand the impacts of the legislation we must also review the WQA in its entirety. The purpose of this chapter is to highlight the more significant differences between the CWA and the WQA. The following sections of this chapter address the WQA title by title, and describe the major concepts captured in the WQA (leveraging, primacy). We then offer a comparison of the major differences between the two Acts.

The Major Components of the Water Quality Act

The WQA is divided into six separate major elements, or titles. Five of these titles address the same issues as their counterparts in the CWA, although the specifics of many of these titles and sections were amended as part of the development of the WQA. The additional title added to the WQA, Title VI, forms the basis of the CWSRF program, and represents the single biggest change from the CWA. The addition of Title VI to the WQA, and the retention of Title II (the construction grants program), meant that the WQA offered two major federal programs from which to fund clean water needs. This section of the chapter will address each title of the Act individually; our emphasis will be on the parts of the WQA that represent a major shift from the CWA.

Title I

The first title of the WQA contains authorization amounts for several planning, research, and scholarship programs found in this title and others. The authorizations provide funding authority through FY 1990. Title I also empowers the EPA to set aside up to $1 million per year to fund a small flows clearinghouse to collect and disseminate information on small flows of sewage and innovative or alternative wastewater treatment techniques and processes. Section 105 empowers the EPA to conduct research into the effects of pollutants on the health and welfare of humans.

The other two elements of Title I are geographically specific programs to address pollution issues in the Chesapeake Bay watershed (Section 103) and in the Great Lakes region (Section 104). For the Chesapeake Bay Program, Congress authorized $3 million annually through FY 1990 to collect, coordinate, and disseminate research on the environmental quality of the watershed. In addition, the Act authorizes $10 million per year through FY 1990 to provide 50 percent planning grants to the states within the watershed to implement an interstate management plan for the watershed. The Great Lakes program is similar, although the purpose is to meet the requirements of the international agreement with Canada (the Great Lakes Water Quality Agreement of 1978). Under this section, EPA is to conduct a five-year study of toxic substances in the Great Lakes, with particular focus on five different sites. Also included is an authorization of $11 million per year through FY 1991.

Title II: The Construction Grants Program

One of the longer and more detailed titles in the WQA, Title II represents the original funding mechanism for the CWA of 1972. A number of the sections address smaller programs or set certain conditions on process. For example, Section 201 provides a mechanism to resolve disputes, and Section 202 limits the federal share of Title II funds to 75 percent of project costs. It is notable that Section 202 also names several projects in specific states (in Pennsylvania and Minnesota) that may receive more than the 75 percent share. Sections 208 and 209 provide funding for innovative and alternative projects (Section 208) and funding for regional organizations (Section 209). Section 210 provides funding for estuary studies and marine combined sewer overflows (CSOs); Section 214 provides a specific exception to allow funding for the Chicago (IL) Tunnel and Reservoir Project; and Section 215 provides an exception to allow the City of Nashua, NH, to continue to use its ad valorem user charge system to collect the costs of operation and maintenance for their sewage treatment works. Section 213 also includes funding for specific wastewater treatment projects in several communities around the nation, including projects in California, Pennsylvania, Kentucky, New Jersey, Illinois, and Wyoming.

Several other sections address technical processes related to the construction grants model. Section 203 requires that, prior to the award of a grant, there must be agreement between the local government, the state, and the EPA on the eligible costs to be covered under the grant. The section also provides for an audit process for eligible costs, including the ability to recover funds for unallowable costs. Section 206 provides for the allotment formula to be used to allocate construction grant funds to states, and creates an extension for the minimum allotments for each state through FY 1990. This section also continues the set-aside provision for state management of the construction grants program, and extends a prohibition against the funding of separate storm sewer systems through FY 1990.

Section 205 requires states to implement areawide waste treatment management plans, and requires any funded projects in those areas to be included in the plans. This section also provides an exception to allow lower proportional shares of sewer treatment charges for low-income residential users. This is one of several elements of the WQA that indicates congressional intent for the WQA to serve low-income residents or, as is the case with other requirements, to serve communities facing financial hardship. Similarly, Section 207 creates a set-aside in the program to serve rural communities. States with more than 25 percent rural population are required to set aside a minimum of 4 percent of their funds to serve rural populations; at the request of the state's governor, this set-aside may be increased to as much as 7.5 percent.

Section 212 amends the WQA to include the new Title VI, which creates the Water Pollution Control Revolving Loan Fund program. The specifics of this new title are discussed later in this chapter.

Title III: Standards and Enforcement

Title III specifies standards and enforcement procedures under the WQA. In broad terms, this title provides legal authority for the EPA to collect and analyze data on potential pollutants, and to regulate those pollutants. Section 301 sets new dates for compliance with the requirements for priority pollutants, and for meeting "best available technology" (BAT) requirements. The BAT requirements were some of the more controversial elements of the CWA, and amendments to the CWA in the years between 1972 and 1987 pushed the dates for compliance back – a process reiterated here. Section 302 specifies five "nonconventional pollutants" (ammonia, chlorine, color, iron, and total phenols) that may be exempted from the BAT requirements. This section also provides for a mechanism for other pollutants to be added to the list for a similar exemption.

Section 303 addresses the discharge of pollutants into marine waters. The language in this section expressly forbids the discharge of pollutants into marine waters that would interfere with the attainment or maintenance of water quality standards for that body of water. In addition, treatment works serving populations of 50,000 or more must either pretreat effluent or provide appropriate secondary treatment to remove toxic pollutants, suspended solids, and at least 30 percent of the biological oxygen demanding (BOD) material, and provide for disinfection where appropriate. The language in the section also requires all future permits to meet these standards, but also expressly forbids modification to existing permits for discharges into the New York Bight Apex.

This section also contains what has become one of the more controversial elements of the CWA. As part of the process to control the discharge of effluent into marine waters, the CWA creates a process through which states identify impaired waters and create plans to limit the discharge of pollutants into these waters. Located in Section 303(d) and called the Total Maximum Daily Load

(TMDL) process, the underlying logic is that each impaired waterway can be restored by limiting the amount of pollutants found in the waterway. By setting TMDL limits, states can improve water quality in these waters by imposing stricter limits on discharges. More importantly, if states do not create and implement these limits, the EPA is empowered to impose TMDL limits. However, as Houck (2002) points out, the states did not act on this provision, and neither did the EPA. Following the passage of the WQA, which retained this provision, a spate of lawsuits in the 1980s and 1990s compelled states to both list their impaired waterways and develop TMDL plans. While a full analysis of Section 303(d) is beyond the scope of this book,[2] the TMDL process remains an important and controversial element of the WQA, and court battles and implementation issues continue to the present day.

Sections 304 and 305 address technical issues regarding deadlines for filing modifications for treatment works and for treatment plants that utilize innovative technology. Section 306 empowers the EPA administrator to create alternative requirements and standards for treatment plants that can demonstrate they are fundamentally different than other treatment plants in the industry. The purpose of this section is to allow for innovative technologies to be considered separately from more conventional treatment processes. The section also provides for the collection of fees from applicants under this section (and Section 316) to offset the costs of data review and processing. Section 307 allows for the use of "Best Professional Judgment" (BPJ) in the case of effluent from coal remining operations. Under this section, EPA must consider discharges of iron, manganese, and changes in pH on a case-by-case basis. However, the levels may not exceed those previously approved, and the applicant must demonstrate that the remaining operation provides the potential for improvements in water quality.

Section 308 provides a process to address individual control strategies for toxic pollutants. While the requirements of Section 303(d) tend to focus more on nonpoint-pollution sources, Section 308 specifically addresses point-source pollution. Under this section, states must develop plans for the control of toxic pollutants from point sources in specific waterways. These plans were to be submitted to EPA no later than February 1989. These plans contain not only information on the amounts of specific pollutants covered in this section, but also methodologies to measure water quality criteria. After approval by EPA, states would have three years to implement those plans. The EPA Administrator must also publish a biennial plan to review and revise the guidelines, identify new pollutants not already covered, and establish a timeline for any new pollutants added.

Sections 309–313 address fairly narrow technical issues. Section 309 amends Section 307 by extending the compliance date for the use of innovative technologies for up to two years, assuming the extension will not result in a violation of the plant's discharge permit, and if the EPA Administrator determines that the new technology has the potential for industry-wide application. Section 310 provides for penalties for any government official who discloses confidential information. An important addition to this section is the inclusion of authorized EPA contractors as "authorized representatives" of the administrator. However, Section 310 also requires the administrator to divulge any information to Congress upon written request. Section 311 addresses marine sanitation devices on vessels. This section empowers states to adopt regulations that are more stringent than federal requirements for houseboats that are used as primary residences (and not primarily used for transportation). States are also authorized to enforce federal standards for all vessels, regardless of primary use. Formerly the purview of the US Coast Guard, this provision allows states to enforce the use and discharge of marine sanitation devices in state waters. Section 312 expands criminal penalties for violations of the provisions of the Act. The language in the section differentiates between negligence and intent, and ascribes significantly higher penalties for crimes of intent (doubled fines and up to three years in prison). This section also specifically defines false statements as a criminal act, and prescribes both fines and jail terms for this offense. Finally, Section 313 provides for fines of up to $25,000 per day per violation for violators of the National Pollution Discharge Elimination System (NPDES) permit or pretreatment requirements found in Title IV. This represents a civil penalty, but the language in the WQA increases the fine significantly from its previous level of $10,000 per day per violation. The section also requires states to match the minimum penalty imposed by EPA for each violation.

Section 314 addresses administrative penalties under the WQA. The EPA may issue civil administrative penalties for certain violations of the WQA, or for violations of NPDES permit conditions. The law creates two different classes of administrative violation. Class I violations carry higher fines (of up to $25,000 per violation), and allows the person charged with a violation to request a hearing. Class II violations carry lower fines (up to $10,000 per day per violation), and also caps total fines at $125,000 per violation. A person found guilty and assessed a Class II penalty has the right of appeal in federal district court; a person appealing a Class I penalty does so in a Court of Appeals. This section also specifies that if EPA is actively pursuing a prosecution with an administrative penalty, the violation may not be subject to a civil judicial penalty or to a citizen suit. Citizen suits may only be allowed if filed (or notice of intent to sue is given) before the administrative proceedings begin.

Section 315 requires states to submit a biennial report that addresses water quality in lakes. The report includes a description of the lakes, and information about water quality issues in those lakes, and details state efforts to address water quality in the lakes, control sources of pollution, and address issues related to acidity. This information is compiled by EPA and sent to Congress in the form of a national report; the EPA must also develop and disseminate a lake restoration guidance manual. The section also authorizes a total of $55 million for demonstration projects, but states that fail to report as directed under this section are not eligible to receive demonstration grant funds. The section also details requirements for demonstration projects in several states.

Sections 317 and 320 create a national estuary program. Governors of states may nominate an "estuary of national significance" to receive funding for a management and planning conference. The goal of the conferences is to develop a comprehensive conservation and management plan to address priority actions and compliance targets. These conferences may be convened for up to five years. The EPA will provide grants of up to 75 percent of the costs of these conferences, with the balance to be covered by the participating states. The sections also authorize $12 million per year for FY 1987–1991 for this program. Section 320 also requires priority consideration to be given to eleven different estuaries; notably missing from this list is the Chesapeake Bay estuary, although the Chesapeake Bay Program is addressed under a separate program under Title I.

Perhaps the most important element of Title III is the addition of Section 319,[3] which addresses nonpoint-pollution programs. This section creates a national program for the management of nonpoint sources of pollution. Each state is required to prepare a report that identifies state waters which, absent any additional action to address nonpoint pollution, would not be expected to attain or maintain applicable water quality standards pursuant to the WQA. The reports are required to provide detailed information regarding the categories and sources of nonpoint pollution, and to identify state management processes and pollution control programs relevant to nonpoint pollution. States are also required to submit a management program to control nonpoint pollution; the report must include specific control programs, implementation schedules, funding sources, and legal authorities. States are encouraged to develop these plans to cover entire watersheds, rather than portions of watersheds, and to engage local, substate, regional, and interstate authorities in the plans. Upon submission of the reports, states are eligible to apply for EPA grants of up to 60 percent of management costs to assist with the administration of the management program. Construction and cost-sharing expenses are addressed separately; states may also apply for grants to engage in groundwater protection at a rate of up to a 50 percent federal share. Section 319 authorizes up to $400 million over four years to fund these management programs; an additional $7.5 million is authorized for the ground-water protection management plans. Finally, the law provides for a priority system in the grant award process that focuses on innovative practices, particularly impaired waters, interstate problems, or programs that are part of a comprehensive nonpoint pollution control program.

Title IV: Permits and Licenses

Title IV of the Water Quality Act contains the bulk of the regulatory requirements in the Act, and builds from or modifies the programs set forth in the Clean Water Act. At the heart of the regulatory regime is the NPDES, the name given to the permitting process for water pollution discharge. The NPDES has been analyzed in detail by others (see, for example, Lowry 1992; Hoornbeek 2011); our purpose here is simply to outline the nature of the program. In short, all point sources of water pollution must apply for, and be granted, a permit to discharge effluent. The permit is issued by the EPA, but under Section 403 states can apply for partial delegation authority to both issue and monitor permittees within the state. Section 404 also requires that no permit can be issued that leads to a less stringent water quality standard. Section 405 sets new standards and procedures for the permitting of municipal and industrial discharges. For municipal and industrial systems in areas with more than 250,000 citizens, the permittee is required to be in compliance with its permit no later than seven years from passage of the WQA. For municipal systems serving more than 100,000 residents, the permittee has nine years to attain compliance with the permit. This section also specifically delays requirements for stormwater permits until the beginning of FY 1993, and prohibits the discharge of non-stormwater effluent into storm sewers.

Section 406 addresses the development of criteria for addressing sewage sludge. During the wastewater treatment process, solids are removed from the waste stream, either through a process of filtering/screening or through the use of coagulants. These solids often contain toxic elements and must be disposed of in a controlled manner. Section 406 requires the EPA to publish regulations regarding the treatment and disposal of sewage sludge no later than June 15, 1988, and requires compliance on the part of treatment plant operators within one calendar year. The processes relating to the treatment and disposal of sewage sludge are now also made part of the NPDES permitting process. Finally, the EPA is charged with creating and updating a list of regulated toxins to be covered under the NPDES process.

Title V: Miscellaneous Provisions

The penultimate title of the WQA contains a number of miscellaneous programs and provisions. For example, Section 502 defines the North Mariana Islands as a "state" for the purposes of the WQA, thus providing legal authority to apply for and receive federal funds under the WQA. Section 503 specifically excludes agricultural stormwater discharges as a form of point-source pollution, which thus removes the requirement for these discharges to be subject to the NPDES process. Section 506 provides additional authority to Native American tribes. Under this section, EPA is required to assist tribes in the planning and construction of treatment works, and tribes are given status as "states" in terms of the WQA. Tribes are also eligible to receive funding up to 100 percent of project costs, thus removing the requirement for a "state" match for these funds. Section 507 expands the definition of a "point source" to include leachate collection systems associated with landfill operations. The final sections of Title V address specific needs in geographical locations. For example, Section 508 prohibits the dumping of sewage sludge in the New York Bight by December 15, 1987; and Section 511 limits the discharge of raw sewage by New York City. Section 510 authorizes grants to the City of San Diego, CA, to address sewage crossing the international border from the city of Tijuana, Mexico. Other sections address projects in specific areas as well, including Boston Harbor, De Moines, Iowa, and treatment plants in the New York Harbor area. The final elements of Title V require EPA to fund a series of studies and demonstration projects to address issues such as the pretreatment of toxics (Section 519), water pollution in aquifers (Section 520), the effects of sulfide compounds in collection and treatment systems (Section 522), dam water quality (Section 524), and other similar studies.

Title VI: State Water Pollution Control Revolving Funds

Perhaps the single biggest difference between the Clean Water Act and the Water Quality Act was the addition of Title VI to the 1987 legislation. By altering the means by which the federal government provided funding for water quality to the states, the WQA had a profound impact on federal/state/local relationships in the water quality arena. The language in Title VI describes a rather complex funding process to transfer grant funds from the national government to state governments; it also spells out state reporting and use requirements. Title VI is subdivided into seven sections; we consider each of these sections in turn.

Section 601

Section 601 provides for delegation of federal authority for the CWSRF program to the EPA Administrator. The administrator is responsible for disbursing the capitalization grants to states, subject to several other provisions of the WQA. Title VI funding employs the same definition of "treatment works" as found in Section 212; Title VI does not attempt to alter the kinds of project for which funding may be used. Section 601 also allows for the use of CWSRF funds by states to implement a management plan as defined by Section 319 (nonpoint-source management programs), and for developing and implementing conservation and management plans as defined in Section 320 (national estuary program).

Section 601 also defines the requirements for a schedule of grant payments to states. The plans are to be jointly established by the administrator and states, and must be based on state Intended Use Plans (IUPs)[4] as defined under Section 606. Section 601 also requires that the payments to states be made on a quarterly basis. This section further requires the administrator to make the payments "as expeditiously as possible, but in no event later than the earlier of" either two years after the funds are obligated by the state; or within three years of appropriation.

Section 602

Section 602 details the requirements and form for the capitalization grant agreements, and specifies the disbursement of funds from the EPA to the states. Under the agreement, the EPA disburses grant funds to the states' CWSRF accounts in quarterly payments. The state must also deposit an amount equal to 20 percent of the federal grant funds into the same account.[5] Once the funds are in the account, the state must enter into binding agreements for all of the funds (federal funds plus the required 20 percent state match) within one year of receipt of the grant funds. The purpose of this provision is to ensure that states did not just "bank" the federal grant funds, but pushed the funds out to communities in the form of loans and other allowable purposes. The intent is to ensure that all federal funds are allocated by states immediately; once the grant funds are revolved back into the fund in terms of loan payments, the one-year requirement is no longer relevant. This section also contains the so-called first-use requirement. The first-use provision requires that "...capitalization grants under this title and section 205 (m) of this Act will first be used to assure maintenance of progress, as determined by the Governor of the State, toward compliance with enforceable deadlines, goals, and requirements of this Act..." (WQA 1987, Section 602(b)(5)). In short, this requirement places significant environmental need as the primary category of need for the initial use of Title VI funds. Finally, the latter elements of Section 602(b) specify the requirements for states to comply with generally accepted governmental accounting methods when reporting fund use under Section 606 (audits and compliance).

Section 603

This section is the longest and most detailed of the seven sections of Title VI. Section 603 details the state requirements to create state programs, defines the types of assistance states may make available (including loan length, payment schedules, and interest rates), discusses funding for administrative costs incurred by states, and links the priority list process to Title VI funds. In short, Section 603 details congressional expectations for the operation and administration of state programs.

The first requirement is for states to establish state legislative authority to create a revolving loan fund to receive the federal capitalization grants. This legislation must also specify a responsible entity within the state to administer the program, and provide that entity with the legal authority to act on behalf of the state and to make decisions consistent with purposes and intent of the program. Section 603(c) is particularly important. The bulk of the language here places limits on the kinds of project that may be funded through the CWSRF (again referencing Sections 212, 319, and 320 as the location of the operative definitions). However, it is in this section where we find the requirement for states to operate the CWSRF program in perpetuity: "The fund shall be established, maintained, and credited with repayments, and the fund balance shall be available in perpetuity for providing such financial assistance" (Title VI, Section 603(c)). By accepting a capitalization grant, states are thus committed to maintaining funds in the account forever.

Section 603(d) lists the types of assistance allowed for the program, and thus limits the ways in which the funds may be used. There are seven allowable uses of CWSRF funds listed in this subsection:[6]

- To make loans for water quality uses. If CWSRF funds are used for loans, the loans must be made at rates equal to, or below, the prevailing market rate; states may also offer interest-free loans. The loans are limited in term to no more than a twenty-year amortization period.[7] The section specifies that loan repayment must begin no later than one year after project completion. The loan recipient must also establish a dedicated source of revenue to repay the loan. All loan payments (principal and interest) must be paid to the state's CWSRF fund.
- CWSRF funds may be used to refinance existing municipal, intermunicipal, and interstate debt at or below market rates, but only for debt incurred after March 7, 1985.
- The state may use CWSRF funds to either provide guarantees for the purchase of insurance, or purchase insurance outright, for local debt in cases in which such action would lead to either reduce interest rates or enhance access to credit markets.
- CWSRF funds may be used to pay for (or secure) principal and interest on either revenue bonds or general obligation bonds, if the proceeds from those bonds are deposited in the CWSRF fund. This provision allows states to meet the 20 percent match requirement in ways that do not involve the direct appropriation of cash from general revenues to the program.
- CWSRF funds may be used to provide loan guarantees for similar state-run revolving fund programs.

- CWSRF funds may earn interest on monies in the fund.
- States may use up to 4 percent of federal capitalization grants for administrative costs. This requirement was a major point of concern for states, many of whom were concerned that administrative costs for the program would exceed the 4 percent cap. By limiting expenses to 4 percent, Congress signaled a clear intent that federal funds should be allocated to direct project costs.

Section 603(e) prevents communities from receiving double benefits for construction planning. Communities may not receive both CWSRF funds and Section 201 (planning grant) funds for the same project. If they do, they must repay the CWSRF funds used for planning promptly. Subsection (f) further defines the projects eligible for funding as those consistent with project plans developed under Sections 205, 208, 303, 319, and 320.

Section 603(g) links CWSRF funding to the priority list process created in Section 216. A remnant of the construction grants program, the original intent of the priority list was to provide for state input into which communities were to receive Title II funds. However, funding decisions for Title II funds were made by the EPA Administrator. Under Title VI the priority list is still compiled in the same manner, but now becomes an accountability mechanism to ensure that communities receiving funds are part of a process that accounts for the environmental priorities of the state. Under Title VI, the priority list becomes the foundation for CWSRF assistance, and its use is reflected in both a state's IUP and its annual report.

Section 604

Section 604 describes the process for the allotment of funds under Title VI. The first element of the section requires that the funds be distributed by EPA in accordance with the funding formula found in Section 205(c) of the legislation.[8] This section also requires states to set aside a minimum of 1 percent of the federal grant funds (or $100,000, whichever is greater) to fund the planning requirements found in Section 205(j) (water quality management grants) and Section 303(e) (development of state planning process) of the Act. The last paragraphs of the section make the appropriated federal grant funds available to states through the fiscal year in which they are allocated, as well as the following year. If states do not have projects ready to fund in a given fiscal year, they can delay receipt of those funds for up to one fiscal year and still receive the full amount. This provision gives states extra latitude to ensure that recipient municipalities are ready to proceed, but also ensures that states act in a timely manner. The final element of Section 604 allows the EPA Administrator to reallocate any funds not claimed by states at the end of the two-year period defined above; these funds are distributed to all eligible states on the basis of the same funding formula used to calculate the amount of the initial grants.

Section 605

Title VI has certain penalties for noncompliance on the part of states. Section 605 provides the EPA Administrator with the authority to withhold payments to states if the administrator determines that the state is not in compliance with the requirements of the WQA. If the administrator withholds funds from a state, the state has twelve months to take the necessary corrective action to once again be in compliance. The penalty for failing to achieve compliance is a permanent loss of that year's funding allocation. If this occurs, the unallocated funds are distributed to eligible states under the percentages found in the funding formula.

Section 606

The penultimate section of Title VI serves, in effect, as the accountability provision of the CWSRF program. In addition to stressing the requirement for states to comply with accounting and auditing standards, this section also describes several important documents required of states under Title VI. The first of these is the IUP. An annual requirement, the IUP identifies how the state plans to employ its Title VI grant for the upcoming fiscal year. The uses described must also be linked to both the goals of the CWSRF and the WQA, which include the short-term and long-term goals of the state program, the kinds of community it intends to serve with the federal funds (i.e., environmental need, small communities, hardship communities), and the terms of the assistance the state intends to provide. The document must also describe the projects to be funded, including the project category and how the project meets the goals listed in Titles III and IV of the WQA. The IUP must be subjected to public comment and review within the state before submission to the EPA, and the IUP must be accepted by EPA before a capitalization grant may be awarded. Finally, the IUP also provides written assurances and proposals for meeting the other specific requirements of Title VI.

Section 606 also details the requirement for an annual report to be prepared by each state and sent to EPA at the beginning of the first fiscal year following receipt of grant payments. The annual report explains how the state met the goals and objectives the WQA with the funds from the previous fiscal year as listed in that year's IUP. The report must provide detailed information about loan recipients, assistance amounts and objectives, loan terms, and other information. This report, along with other information required by EPA, is now captured (in part) in the National Information Management System (NIMS), where all CWSRF funds and project assistance are tracked.

Finally, this section requires the EPA Administrator to conduct an annual oversight review of all state materials required under Section 606, and also gives the administrator the authority to request any additional materials or information "considered necessary and appropriate" for the oversight review. These materials can be requested of the state, or of any loan recipient as needed to ensure compliance. Finally, Section 606(f) draws a clear line between the construction grants program and the CWSRF: "Except to the extent provided in this title, the provisions of title II shall not apply to grants under this title." The intent is to separate the two programs. However, the discerning reader will also note that states that receive both Title II and Title VI funding must thus prepare and submit separate compliance documents to the EPA – one set for each program.

Section 607

The final section of Title VI provides the authorization for the CWSRF program. Authorizations were approved for a period of six years, beginning in FY 1989 and concluding in FY 1994. For FY 1989 and FY 1990, the authorized amount for Title VI was $1.2 billion.[9] The authorized amount rose to $2.4 billion in FY 1991, and then tapered down to the final year of funding: $1.8 billion in FY 1992; $1.2 billion in FY 1993; and $600 million in FY 1994. Because the Title VI funds were intended as "seed money" to states to establish the state programs, the intent of Congress was to end appropriations at the end of the initial period of authorization. Federal investment in the CWSRF program would thus be limited to a total of $7.2 billion over five years.

In reality, this investment limit proved robustly inadequate. Although authorization for the CWSRF expired at the end of FY 1994, Congress has continued to appropriate funds to the CWSRF program every year since FY 1994, in amounts varying from $491 million in FY 1997 to $1.707 billion in FY 2019. The reasons for this continued appropriation are complex, and are addressed in other parts of this book, but the salient point is that the initial authorization found in Section 607 proved inadequate to meet state water quality needs. It is also worth noting again that, despite the ongoing appropriations, Congress has never reauthorized either the WQA or Title VI. Indeed, even Title II, for which authorization expired in FY 1990, continued to receive appropriations through FY 2013.

Summary

The WQA retained much of the underlying structure of water pollution control as envisioned by the CWA, but fundamentally changed the means by which the national government would support the costs of compliance with the water quality standards. By replacing categorical grants with a block grant program, the WQA fulfilled an important goal of Reagan's new federalism – the return of policy authority and, in theory, eventual state assumption of budgetary responsibility for water quality funding. State policy authority was also extended by the inclusion of provisions for partial state delegation in the NPDES process. The WQA retained the permitting and regulatory process, with only minor modifications. The other major change was the enhanced focus on nonpoint pollution, as found in the new Section 319 program. The specifics of these changes will be discussed in more detail later in this chapter; we first turn our attention to the specific processes and concepts resulting from the switch to a different grant funding mechanism.

Leveraging

Leveraging is a critical element of the revolving loan fund model. The concept of leveraging was an integral part of the EPA Task Force report issued in 1984 that developed the revolving loan fund model. Given the realization that federal budget reductions had to be a part of any successful revision to the CWA, the EPA Task Force recommended a model that would, in theory, trade substantial short-term federal investment for the long-term cessation of federal funding for water quality. However, the task force also knew that any capitalization funds likely to receive approval of Congress and the Reagan administration would not be sufficient to capitalize state programs to the point they could become self-sufficient. It was also clear that state budget pressures would not allow states to make additional investments large enough to meet state needs. The task force thus encouraged the use of leveraging – the ability to use federal capitalization grants and/or existing state CWSRF funds as a guarantee to back the issuance of state bonds to provide additional funds with which to make loans to applicant communities.

The legislation does not explicitly mention leveraging, nor does it encourage the use of leveraging. However, Section 603(d) does provide the legal authority necessary for states to apply the federal grant funds to a leveraging arrangement. By allowing states to apply the CWSRF funds in the form of loan guarantees, Congress allows for the possibility of leveraging. Congress (and the EPA Task Force) understood that not all states would either want, or need, to leverage, so the WQA does not require states to engage in leveraging. But, for states with significant water quality needs, leveraging provides a mechanism for states to increase significantly the amount of funds available for loans. If a state chooses to apply an aggressive leveraging model (i.e., to issue bonds in an amount many times the value of the capitalization grant), the state can generate funds equal to six or seven times the value of the capitalization grants. States wishing to take a more conservative approach could also leverage, but limit their leveraging to two to three times the value of the capitalization grant. In this manner, the choice is not only whether to leverage, but the leveraged amount is highly scalable to meet state needs in that year and can thus be tailored explicitly to the number of projects ready to proceed, and the funds needed for those projects.

The general concept of leveraging is reasonably straightforward. The logic of leveraging is that an investor can borrow additional funds to have more money available for investment. This allows for potentially significantly greater funds available in the short term to apply to the investment – in the case of the CWSRF, leveraging increases the amount of money available for loans to recipient communities. In the short term the fund can continue to grow as long as the return on the investment (i.e., the interest charged to recipient communities) exceeds the interest rate paid on the borrowed funds. For example, if a state borrows $100 at a 2 percent interest rate but loans that same $100 at 3 percent, the value of the state fund has increased by 1 percent. However, long-term loans are also affected by the rate of inflation. If the rate of inflation over that same period is 1 percent, then the fund loses the 1 percent gained because inflation has devalued the account by an amount equivalent to that gain. Because CWSRF loans may be made for terms up to thirty years, the loss to inflation can be substantial.

Leveraging also carries significant costs. Because the state issues bonds under the leveraging agreement, the bonds are payable with interest to the bondholder. For the CWSRF fund to be available in perpetuity, the interest paid on those bonds must be recouped to the fund; this is typically accomplished by charging the loan recipient a higher interest rate on the CWSRF loan. The bond issuance and tracking process also carries administrative costs that must be somehow recouped. If these costs are passed on to the borrower, the cost of the CWSRF assistance can grow to the point that it may be less expensive for the community to seek funding through the private market or, if possible, to issue bonds directly. As reported by Heilman and Johnson (1991), the demand for CWSRF assistance in the early years of the program was impacted significantly by the costs associated with leveraging.

In spite of the costs, leveraging is common in the CWSRF program, primarily because it is a means to generate a significant amount of money with which to meet state needs. As we will see later in this volume, the leveraging models (and amounts) employed by states vary widely, but all states that leverage face the same challenge: How to package the CWSRF assistance in such a way that it is attractive to potential borrowers, and yet meet the legislative requirement to maintain the CWSRF fund in perpetuity. There are a number of strategies that states employ to meet this dilemma. Some states have offered loans at greatly reduced interest rates; others have engaged in loan forgiveness programs; still other states have packaged CWSRF bond issues in such a way as to take advantage of higher bond ratings (and thus lower interest rates). Other states have simply subsidized the potential losses through increased state matches. We should also note that the increased (and initially unexpected) federal appropriations to the CWSRF program have allowed states to adopt varying financial strategies to better balance the need for leveraging.

A report issued by the Environmental Financial Advisory Board (EFAB) in 2008 recommended a number of revisions to the CWSRF program, mostly centered around the use of leveraging by states. The report explored ways to make leveraging more attractive to states, and recommended states use a blend of direct loans and leveraging in order to maximize the impact of CWSRF resources. Among other things, the EFAB recommended the use of accelerated capitalization draws rather than tying the draws to construction costs; the exclusion of the CWSRF program from federal arbitrage regulation; the use of innovative investment strategies by states; and the expansion of the maximum loan period from twenty to thirty years. The board also recommended that EPA interpret the "perpetuity rule" on a dynamic, rather than a static, basis. This would allow states to measure compliance based on expected earnings over time, rather than current year-end results (EFAB 2008). The main thrust of the report was to provide additional incentives for states to engage in leveraging. As of this writing, the only recommendation implemented was the change in loan terms to allow loans up to thirty years, which was included in the Water Resources Reform and Development Act (WRRDA) in 2014.

Letter of Credit

The letter of credit (LOC) mechanism was first used in the United States in 1964 as a cash management tool to allow the recipients of federal grants and contracts to access those funds as needed (OMPC 1988). EPA adopted the use of the LOC as an administrative tool to track the distribution of Title VI grant funds to states. The letter of credit process for the CWSRF program was developed within EPA and released in 1989 as part of the administrative rules for the CWSRF program. The genesis of the letter of credit is found in Sections 601 and 602, which require the disbursement of grant funds in quarterly payments. Rather than transfer funds directly from the US Treasury to state treasuries, the letter of credit process involves an interim transfer to a third-party (private) banking institution.[10] The private banking institution is chosen by the state but must be approved by the US Treasury.

Under the letter of credit process, the Office of Management and Budget (OMB) apportions funding authority to the EPA each fiscal year. The funds are held in the US Treasury, but under EPA authority. The EPA is charged with allocating the appropriation for each fiscal year according to the funding formula set out by Congress. When a state is ready to proceed it requests a transfer of funds from

EPA. At this stage, EPA only checks to ensure that the fund balance is accurate, but does not review the programmatic justification for the request. If EPA does not challenge the request, the funds are automatically transferred electronically[11] to the recipient bank. The letter obligates the funds to the state, and the amount is deducted from the state's total appropriation for that fiscal year. Under the requirements of Sections 602 (use of funds within twelve months of receipt), 604 (limit of twenty-four months to request funds), and 605 (ability to withhold funds), the state may then apply the grant funds to any eligible purpose. EPA does not review the programmatic validity of the request until the end-of-year audit process. This audit process ensures that all legal requirements are met (including an approved IUP).

The funds remain in the private banking institution until requested by the recipient community (or the state, if the funds are to be used for leveraging, administrative costs, or other non-loan approved purposes). The state agency contacts the bank, which then requests a transfer from the treasury for the requested amount. The treasury then wires the funds for deposit to the private bank, and the funds are then available within one to two days of the request to the treasury. The funds are then disbursed by the private bank to the designated recipient. The amount drawn against the letter of credit is determined by a formula that takes into account the state match and the federal grant. For a state that only meets the match requirement of 20 percent, the federal share is 83 percent (federal grant / (federal grant + state match) = federal share). If states contribute beyond the minimum required match, the federal share drops proportionally.[12] Funds to be used for state CWSRF administrative expenses (allowable up to 4 percent of the capitalization grant) may be drawn at a rate up to 25 percent of the total administrative cost per calendar quarter.

According to the Office of Municipal Pollution Control (OMPC) (1988), the letter of credit process allows for a phased transfer of funds. Previous experience with the Title II (construction grants) program indicates that the construction period for wastewater plants is typically five to seven years; to pay the entire construction cost to the state in a two-year period would mean than federal outlays would increase significantly (OMPC 1988). To control these outlays, the funds are released as needed to meet ongoing construction costs. The letter of credit process thus draws on a well-understood and well-tested administrative mechanism.

In an early review of the CWRSF program (Heilman and Johnson 1991), the letter of credit process was often cited by state program coordinators as overly burdensome and unwieldy. They noted the additional paperwork involved in the process, along with the reduced control over funds allocated to the state. Under the law states earn interest on the funds in the account, and once the funds have been "revolved" (issued in the form of a loan and recouped through principal and interest payments from loan recipients) the state may place those funds in the institution of its choice (and earn interest on that account as well). However, the letter of credit process does play at least a couple of important roles. First, by keeping the funds in the treasury until they are ready to be spent, the interest available on those funds is captured by the treasury rather than by states. Because capitalization grants are only disbursed in quarterly draws, the interest over the fiscal year is not insubstantial. Second, the letter of credit provides a series of accountability "checks" in the system. This prevents the disbursement of funds to states not prepared to receive funds under the requirements listed in the WQA. It also means that the funds (and any interest earned on those funds) remains with the treasury until the state has recipients ready and able to proceed with their projects. Finally, the requirement to involve a private banking institution furthers the policy goal of a greater role for the private sector in the administration of federal programs. There is little benefit to be gained in terms of program administration to involve another step; indeed, such a step adds both cost and administrative burden. However, the requirement is entirely consistent with the metapolicy of privatization, and the creation of market opportunities in the business of government.

The Role of State Primacy

The concept of primacy represents an important feature of federalism, particularly in the realm of environmental policy. Primacy has been a feature of American federalism for several decades. While primacy can be thought of as an implementation structure (Crotty 1987, 53), it also represents a fundamental redefinition of the relationship between the national government and the states. The ascendancy of primacy in environmental policy in the 1970s and 1980s indicates, on its face, a shift in the relative roles of the national government and the states in the implementation of national policy. At its heart, primacy seeks to provide a greater role for states in policy implementation; the concept is at the heart of several environmental programs, including water and air pollution, drinking water, hazardous waste (see Daley and Garand 2005), and surface mining.

The issue of primacy is tied directly to efforts of the national government to preempt state policy authority in the enforcement of environmental standards (Crotty 1987). Preemption occurs when the national government enacts laws that supersede state laws in the same policy arena. When the Clean Air Act was enacted in 1970, for example, the stated goal was to create national standards that would apply to all states. At that time, many states had no air quality standards, or standards that were so minimal as to be useless (see Rosenbaum 2002; Davis 1998). The trend in American environmental policy over the decades has been one of centralization at the national level; the federal government sets standards that states are required to meet. State may choose to exceed those standards, so federal standards are considered to be minimum requirements. If states fail to comply with national standards, the national government can force compliance.

The purpose of primacy is to provide for more state control over the compliance process. States that assume primacy in a specific environmental program agree to monitor and enforce federal regulations within their state boundaries. States must regularly report to EPA regarding these efforts. If EPA is not satisfied with a state's monitoring and compliance, EPA can revoke that state's regulatory authority and reassert federal regulatory authority, although such revocation is relatively rare (Shimshack 2014). In the early years of these provisions some states, generally states that were forward-leaning in environmental protection (see Lester 1989; Vig and Kraft 2019), were quick to assume primacy, while other states were content to allow the federal government to engage in monitoring and enforcement. In essence, the "laggard" states were content to deflect criticism and complaints regarding these regulatory requirements to the national government. Under the WQA, states were encouraged to assume primacy for wastewater enforcement. Although a number of states were slow to create the administrative and legal regimes to assume responsibility, all but three states[13] now exercise primacy under the WQA.

Comparing the Water Quality Act to the Clean Water Act of 1972

As detailed in the preceding sections, the WQA includes provisions for the cessation of the construction grants program, and the establishment of the CWSRF program. In many respects, this represents the largest change in what is otherwise largely an incremental adjustment to the Clean Water Act. As detailed in Chapters 3 and 4 of this book, the history of water quality policy can largely be described in incremental terms, with occasional substantive change. Just as the formal assumption of federal water quality standards (and the creation of the NPDES permitting process) were substantial changes to the Clean Water Act, so is the inclusion of Title VI the "headline story" of the WQA.

The importance of the CWSRF model in water quality policy cannot be overstated; it represents nothing short of a fundamental redefinition of the relationship between the national government and the states, and it places significant new burdens, both administrative and financial, on states as a result. The impacts of these new burdens, and the abilities of states to assume responsibility for these burdens, are the focus of this book. The WQA simultaneously gives states new freedom of action and a greater role in the provision of clean water, but it still holds states to the requirements of a federal water quality standard. By expanding the nonpoint-source provisions of the Act, Congress is also both addressing an unmet need and increasing the regulatory requirements on states. How states responded to the new framework, and whether the states were able to create, implement, and administer state programs that could meet the legislative intent of the national policy are questions of no small importance. An underlying assumption of the New Federalism is that states would (and could) respond to these challenges; the WQA thus provides an empirical test of those assumptions.

The cessation of the construction grants program represents a substantial policy retreat for the national government. Although the withdrawal from a categorical grant model took nearly five times as long as envisioned by the framers of the WQA, the surrender of national policy authority as exercised in the construction grants is a reasonably rare occurrence in federalism. That the revolving loan fund model has not only survived for more than thirty years, but has also been adopted in other policy areas (e.g., drinking water, transportation), is evidence of the popularity of the model. The continued appropriation of federal funds to capitalize the state programs is not so much an affirmation of the model as it is a realization that, even with leveraging, the estimates of resources needed to allow state compliance with water quality standards fall significantly short of actual requirements. In the following chapters, we will examine the success of the CWSRF program in terms of meeting national water quality standards, and the intent of Congress in terms of the types of community served in the CWSRF program.

Under the CWA, states played a relatively benign role in the construction of water quality infrastructure. States worked with municipalities to prepare construction plans and applications for construction grants, prepared and maintained a statewide priority list for funding, provided state funding support for construction, and monitored the construction process. For states that assumed primacy in the regulatory framework, state agencies also took the lead role in enforcement and compliance. The state legal authority required for these tasks was minimal, and the skills required to complete these tasks were well-understood and available. The funding decisions, approval of projects, review of plans, auditing, and enforcement were carried out by federal actors in either federal regional offices or in headquarters (or both). This arrangement also meant that the planning and growth management functions were retained by municipal governments; since funding decisions were made at the national level, states had little input into the relative impacts on growth tied to water treatment.

The WQA explicitly places the burden for program design, implementation, and administration squarely on states, and relegates the federal role to one of consultation, monitoring, and oversight. The burden lost by the EPA is instead transferred to states. But, because the new funding mechanism is a loan program (that must be maintained in perpetuity), the administrative burden on states is increased substantially. States are still responsible for the same tasks for which they had responsibility under the CWA, but they also assume responsibility for a complicated financial system. Leveraging adds substantial complexity to the system.

The first step for a state legislature was to pass enabling legislation that created the state program, and provided the legal authorities needed to administer the program. Issues such as the agency primarily charged with implementation had to be settled. Many states initially decided to build on existing state capabilities in environmental (or health) agencies, but other states located the program in new or existing infrastructure finance agencies. Many states designated a lead agency, but delegated program authority across three or more agencies, which in turn required specific delegations of authority and interagency agreements for implementation (see Heilman and Johnson 1991). Decisions about the source and nature of state matching funds also needed to be settled; states could choose to meet the match requirements through general revenues, bond sales (within the CWSRF or from without), leveraging, or by redirecting funds from other state water quality programs. Finally, states needed to disseminate information on their new programs to potential loan recipients throughout the state.

An important element to the CWSRF model is the inclusion of a substantial financial component that was not present in the construction grants program. States must now set loan terms (interest rates and loan duration), and create mechanisms to track cash flow. All of this must be done within the context of the requirement to maintain the fund in perpetuity. For states that decide to leverage their capitalization grants, the financial elements of the program become yet more complicated. While some states had preexisting capacity within the state to design, implement, and administer these complex financial programs, other states struggled to secure the financial expertise needed (Heilman and Johnson 1991). A common solution for states was to seek expertise from the private sector (see Morris 1999a), which changed the ways in which loan resources were distributed within the state (see Morris 1997).

As noted earlier in this chapter, the changes in the WQA brought about by the increased emphasis on nonpoint pollution also created new challenges for state governments. The requirements of Section 319 place new information gathering, analysis, and reporting requirements on states, thus increasing the workload on state agencies. In a more direct programmatic sense, the focus on watershed-wide nonpoint plans also required state agencies to work with different stakeholders in each watershed to facilitate the planning process. Although states (and EPA) largely ignored the nonpoint program requirements until pressure from environmental groups and the courts compelled states to act (Houck 2002), the delay was, at best, a temporary reprieve. The nonpoint program in the WQA ultimately proved to be an important new addition to clean water policy; for the first time, states were required to address nonpoint pollution directly. Congress signaled the importance of the nonpoint program by authorizing $400 million over four years to support state efforts; that sum is the second-largest[14] nonconstruction authorization in the Water Quality Act.

Glicksman and Batzel (2010) note that although Congress made a clear distinction between point-source and nonpoint-source pollution, the focus of the Clean Water Act and the Water Quality Act is clearly on point-source pollution. They note that the core provisions and requirements of the WQA apply almost exclusively to point-source pollution; in fact, they note that both the CWA and the WQA fail to provide a specific definition of nonpoint-source pollution. There are no provisions for the creation of standards for nonpoint pollution, and no federal enforcement mechanism – standards and enforcement are left entirely to state discretion. Although Section 319 does require the development of nonpoint-pollution plans, there are no provisions for the development or enforcement of national standards. Similarly, Lowry (1992) argues that the nonpoint program in the WQA was, in effect, an afterthought to the law. While nonpoint programs have been a part of water quality legislation since 1972, the early iterations of these programs were significantly underfunded. The WQA requires states to submit plans to address nonpoint-source pollution; states that fail to submit plans would have plans written for them by EPA. The law also authorizes grants to states of up to 60 percent of the cost of state management plans. However, as noted by Lowry (1992), a total authorization of about $400 million (out of more than $18 billion authorized in the WQA) does not indicate a strong congressional desire to address nonpoint-source pollution.

Lowry (1992) suggests that the lack of attention to nonpoint-source pollution is more easily understood in political terms. Point-source pollution is much more visible, he argues; when combined with the possibilities of pork barrel politics for large infrastructure projects (such as wastewater treatment plants), the benefits of programs for point-source projects over nonpoint projects become clear. Furthermore, he notes, the politically powerful agriculture lobby is opposed to nonpoint pollution control, believing that the bulk of the costs will fall on farmers. Taken together, there is little political benefit to a large nonpoint program.[15] Houck (2002, 5) notes that EPA was hesitant to enforce nonpoint regulations (including the development of TMDL requirements) until forced to take action as a result of several court cases in the late 1980s and 1990s.

Summary

The CWSRF program represents a significant change in the federal government's approach to water quality funding. The shift from categorical grants to a block grant is not in itself a new policy tool, but when combined with the requirement to create a revolving loan fund in perpetuity the change in construction funding becomes much greater in magnitude. The increased focus and funding directed at nonpoint-source pollution also represents a more significant change in the WQA. Most of the policy tools (program grants, reports, earmarks for specific projects, the NPDES permitting process, etc.), however, are holdovers from previous iterations of the clean water legislative program; the inclusion of a significant authorization for the construction grants program also supports a conclusion of what is largely incremental change in the WQA. Just as the 1972 legislation kept the underlying logic of previous water quality legislation intact, the expansion in the scope of the construction grants program, the permitting process, and the "fishable and swimmable" goal add a layer of analytic complexity.

The new funding program enshrined in the Water Quality Act raises questions regarding the success of the new approach to water quality. Were the states able to assume responsibility for the new program, and were they willing and able to address water quality in the ways envisioned by Congress? Has the CWSRF shifted the burden of water quality funding from the national government to the states? Has the WQA had the desired impact on nonpoint-source pollution? And, do we continue to see improvement in the nation's water quality? These questions, and others, are the focus of the analytical chapters found later in this book.

6

A Model of State Implementation of the Clean Water State Revolving Fund Program

In many ways, the Clean Water State Revolving Fund (CWSRF) program is the centerpiece of the Water Quality Act (WQA), and the most significant point of departure from previous water quality legislation. By fundamentally altering the relationship between states, localities, and the national government by shifting to a block grant[1] for water quality funding, Congress redefined the landscape of water quality policy. As detailed in Chapters 4 and 5, the amendments to the Clean Water Act reflected in the WQA were largely incremental in most respects, but the creation of the CWSRF program (and the new Title VI to replace the construction grants program) allowed for a significantly increased role for states in the allocation of water quality resources.

Previous research (see Morris 1997, 1994; Mullin and Daley 2018) indicates that the distribution patterns have been different across the fifty states, but this begs an important question: Why do states make the allocation decisions they do? The purpose of this chapter is to identify a model that explains distribution patterns of program resources, measured as both the dollar value of loans and the percentage of loans made for different purposes. To address this question, this chapter presents a model of state policy choice in the CWSRF program. We will draw on existing models of state policy choice in the CWSRF program, as well as additional theoretical elements from the rather extensive literature on state policy choice, coupled with a review of the existing models of state choice in the CWSRF program. We will then review relevant concepts found in the broader state comparative policy literature, and then present a model of state choice for analysis.

Existing Models of State Choice in the CWSRF Program

A review of the literature in the water quality arena reveals a limited amount of empirical study of water quality resources, little work that employs fifty-state models of policy behavior, and even less work that examines longitudinal data.

A somewhat larger body of work employs qualitative case study approaches of water quality programs, and an even larger body of work focuses on legal or broad policy reviews of water quality programs. Together this literature forms the extant knowledge on the CWSRF program and water quality, and the following discussion assesses the models and data used in that literature to inform the construction of a model for testing in this work. It should also be noted that few existing studies seek explicitly to examine state policy choice in the CWSRF program; to this end, the framework employed to organize the existing literature is somewhat different than the frame developed for this study.

The dependent variables employed in the extant literature range across a spectrum of program elements. A limited body of work focuses on program funds spent. Morris (1994, 1997), Morris et al. (2021), Williamson, Morris, and Fisk (2021), and Clark and Whitford (2011) all employ some measure of dollars spent by states. Morris (1994, 1997) also employs a measure of the percentage of all loans made by each state for certain categories of loan recipients (e.g., small communities, financially at-risk communities). Mullin and Daley (2018, 632) employ a measure of total Environmental Protection Agency (EPA) spending, and a second dependent variable that measures total state-level spending for environmental and natural resource programs. Williamson, Morris, and Fisk (2021) and Morris et al. (2021) use a measure of total CWSRF spending by state (federal capitalization grant + required state match + any additional state spending[2]). Clark and Whitford (2011) use a measure of federal EPA spending and state environmental spending.

A second body of work examines the decision to leverage state CWSRF funds. Travis, Morris, and Morris (2004) explain leveraging as a two-step process, and thus create two dependent variables: the decision to leverage (Y/N); and the amount to be leveraged in dollars. Holcombe (1992) produces an analysis detailing the long-term effects of leveraging and interest rate structures on CWSRF program sustainability, although this work does not take an empirical comparative state approach. A number of other studies (Heilman and Johnson 1991; Morris 1994, 1996, 1997; Bunch 2008) employ leveraging as an independent or control variable.

A third body of work consists of studies that address other elements of state water quality policy. For example, Hoornbeek (2005) employs a fifty-state model to explain nonpoint-source pollution policies across the states. Morris (1999b) examines the distributive effects of targeting of CWSRF resources in the formative years of the program, and Morris (1996) analyzes the involvement of different stakeholders in the development, design, and implementation of state CWSRF programs. Hoornbeek (2011) explains state differences in the stringency of state-issued permits (Whole Effluent Toxicity), and Travis and Morris (2003) and Morris (1997) examine the effects of private sector involvement in the management and administration of the CWSRF program. Ringquist (1993) employs a measure of the strength of state water control programs as the dependent variable.

Independent Variables in Studies of Water Quality Policy

The extant literature on water quality policy includes more than forty independent variables to explain the dependent variables discussed in the previous section of this chapter. There are differences in the ways in which variables are defined and different data sources employed to measure these variables. It is often the case that two studies will use similar measures or, in the case of a control variable, use a different value of a categorical variable as the baseline value. For example, political culture is used in several studies, but while one study might control for states with individualistic cultures (e.g., Hoornbeek 2005), another might control for states with traditionalistic cultures (e.g., Breaux et al. 2010). Moreover, there is little attempt in the extant literature to differentiate between independent variables and control variables, so this review treats all explanatory variables as independent variables. The variables can be grouped into ten different conceptual groups (see Table 6.1); these groups are presented in the following subsections.

Political Explanations

The use of political variables seeks to explain policy outcomes as a result of the interaction of political actors in the policy arena. For example, a common assumption, often supported in the literature, is that the Democratic Party tends to favor policies that seek to enhance environmental protection, while the Republican Party tends to favor environmental deregulation. This concept can sometimes include party control of governor's mansion in the state (e.g., Mullin and Daley 2018; Daley, Mullin, and Rubado 2013; Hoornbeek 2005), but is more commonly measured by employing some measure of party control of the state legislature (e.g., Clark and Whitford 2011; Morris et al. 2021). A similar measure often employed in state comparative studies is a measure of the level of competition between political parties (see Coleman Battista and Richman 2011; Cox, Kousser, and McCubbins 2010; Schlesinger 1955). The logic of this measure is that as competition between parties increases, policy outputs tend to be more balanced. A related conceptualization tests for the effect of party control at the national level. Morris et al. (2021) note that state policy choices also tend to be influenced by party control in Congress and the White House; their longitudinal study of CWSRF spending indicates that state spending is influenced by party control at the national level.

Table 6.1 *Conceptual groups of independent variables*

Group	Variable	Authors
Political	State party control	Mullin and Daley (2018); Travis and Morris (2003); Hoornbeek (2011, 2005); Clark and Whitford (2011); Morris et al. (2021); Williamson, Morris, and Fisk (2021); Breaux et al. (2010); Ringquist (1993); Daley, Mullin, and Rubado (2013)
	Citizen ideology	Travis and Morris (2003); Breaux et al. (2010); Ringquist (1993)
	Government ideology	Clark and Whitford (2011); Williamson, Morris, and Fisk (2021); Morris et al. (2021)
	Political culture	Travis and Morris (2003); Hoornbeek (2011, 2005); Travis, Morris, and Morris (2004); Morris (1994); Morris et al. (2021); Williamson, Morris, and Fisk (2021); Breaux et al. (2010); Woods (2005)
	Federal government party control	Morris et al. (2021)
	Legislative professionalism	Hoornbeek (2011, 2005); Williamson, Morris, and Fisk (2021); Ringquist (1993)
Structural	EPA region	Mullin and Daley (2018); Clark and Whitford (2011); Woods (2005)
	Coastal state	Hoornbeek (2011, 2005)
	% federally owned lands	Hoornbeek (2011, 2005)
	State centralization	Williamson, Morris, and Fisk (2021)
	Land area	Clark and Whitford (2011)
Economic	Gross State Product	Mullin and Daley (2018)
	Fiscal comfort	Travis and Morris (2003); Breaux et al. (2010)
	Per capita income	Clark and Whitford (2011); Hoornbeek (2011, 2005); Morris et al. (2021); Breaux et al. (2010); Ringquist (1993)
	Ability to pay	Johnson (1995)
Sociodemographic	State population	Mullin and Daley (2017); Clark and Whitford (2011); Hoornbeek (2011, 2005); Morris et al. (2021); Williamson, Morris, and Fisk (2021)
	Population density	Morris et al. (2021); Williamson, Morris, and Fisk (2021); Fowler and Birdsall (2020)
Needs	WW needs	Mullin and Daley (2018); Travis, Morris, and Morris (2004); Morris et al. (2021); Morris (1994); Williamson, Morris, and Fisk (2021); Breaux et al. (2010)
	% impaired state waters	Hoornbeek (2011, 2005)
	Environmental need	Johnson (1995)

Table 6.1 (*cont.*)

Group	Variable	Authors
	Demand for loans	Heilman and Johnson (1991); Travis, Morris, and Morris (2004)
Administrative factors	Management experience	Johnson (1995)
	Institutional capacity	Travis, Morris, and Morris (2004); Breaux et al. (2010)
	Initial state sufficiency	Heilman and Johnson (1991); Morris (1999a)
Environmental commitment	State commitment to environment	Travis, Morris, and Morris (2004); Woods (2005)
	State commitment to clean water	Travis, Morris, and Morris (2004)
	State water quality grant programs	Bunch (2008)
	Primacy	Clark and Whitford (2011)
	NPDES authorizations	Hoornbeek (2011)
	NPS activism score	Hoornbeek (2011)
Interest group influence	GSP manufacturing	Clark and Whitford (2011)
	Environmental group strength	Hoornbeek (2011, 2005); Breaux et al. (2010); Ringquist (1993); Woods (2005)
	Industry strength	Hoornbeek (2011, 2005); Breaux et al. (2010); Ringquist (1993)
Program structure/ state choice	Actors involved	Morris (1994, 1996)
	Agency/program location	Travis and Morris (2003); Morris et al. (2021); Williamson, Morris, and Fisk (2021); Hoornbeek (2011, 2005); Breaux et al. (2010)
	Targeting policy	Morris (1994, 1999b)
	Privatization in CWSRF	Morris (1994, 1997, 1999a)
	Interest rate structure	Bunch (2008); Heilman and Johnson (1991)
	Leveraging	Heilman and Johnson (1991); Morris (1994, 1996, 1997, 1999a); Travis, Morris, and Morris (2004); Bunch (2008)
	SRF transfers (CWSRF-DWSRF)	Mullin and Daley (2018)
	Legislation to leverage	Travis, Morris, and Morris (2004)
Misc.	Total federal appropriation	Morris (1994, 1997); Travis, Morris, and Morris (2004)

A second element of political explanations is that of political ideology. There are two measures of ideology typically employed in state comparative studies. The first of these captures the ideology of citizens in a state (Clark and Whitford 2010; Travis and Morris 2003). Citizen ideology is a means to measure policy preferences among citizens in a state, with the notion that the ideology of the active electorate will influence elected policy makers, and thus result in policy outputs that are consistent with that ideology. In this formulation, citizen ideology drives elected officials to be responsive to that ideology, particularly in more liberal states, to favor environmentally friendly policy (Erikson, Wright, and McIver 1989; see also Berry et al. 2010). The second measure of ideology captures the ideology of elected officials (the governor, and party members from each House of the legislature; see Berry et al. 1998). The argument for government ideology, as the latter of these two formulations is known, is that the ideology of policy makers is both more interesting and more relevant, since there is no requirement for elected officials to be responsive in all respects to the ideological positions of what is often an ideologically split citizenry (see Davis and Davis 1999). In other words, elected officials have a certain freedom to act as they see fit,[3] safe in the knowledge that at least some members of their constituency will support that position.

When applied to previous studies of water quality policy, measures of citizen ideology are generally not significant in predictive models (see Travis and Morris 2003; Clark and Whitford 2010), while measures of government ideology tend to be significant predictors (Clark and Whitford 2010; Morris et al. 2021; Williamson, Morris, and Fisk 2021). While a citizen preference for clean water would make sense as a more liberal position, it is also likely that the idea of clean water is not anathema to more conservative citizens as well. Where the two groups differ is in terms of the policies employed to achieve that goal. In the universe of environmental policies, programs to fund clean water infrastructure, such as the CWSRF, are likely to be of relatively low salience to most citizens. On the other hand, because of the sums of money attached to the CWSRF, it makes more sense that, given the absence of a citizen-expressed preference, the ideology of elected officials is more relevant.

A third political indicator often applied in water quality studies is that of political culture. The stream of work that led to the development of political culture as an explanatory variable was begun by Banfield and Wilson (1963). Noting that different cities in different states tended to adopt different policy choices, their work led to the concept of "public-regardingness." A reflection of a state's political/social environment, Banfield and Wilson (1963) argued that cities differed in terms of how city governments adopted policies. While some cities favored policy choices that benefited large numbers of citizens, other cities seemed to gravitate toward policy choices that tended to benefit small groups within the city. Searching for a way to describe and explain similar observations at the state level, Daniel Elazar (1966) developed the term "political culture." Elazar suggested that observed differences in state approaches to policy choices could be traced to historical patterns of migration of different religious and cultural groups across the United States, and that the differences in values and outlook between these groups explained differences in state-level policy preferences.

While Elazar's work has been critiqued over the years (see, e.g., Clynch 1972; Schlitz and Rainey 1978; Savage 1981; Formisano 2001; Chamberlain 2013), the concept of political culture has been employed successfully in numerous studies as a means to understand state-level policy choices. This body of work includes numerous studies of clean water policy, and in most cases the political culture variable has been a significant predictor of state choice in water quality policy (see Hoornbeek 2005, 2011; Travis and Morris 2003; Travis, Morris, and Morris 2004; Morris 1997; Morris et al. 2021). Formulations of the use of political culture vary, with differences reported between traditionalistic states and others, and between individualistic states and others, depending on the dependent variable under study.

Finally, a related political variable found in the literature is that of legislative professionalism. Drawn from the work of Grumm (1971), measures of legislative professionalism typically include elements of the time demands of service, salary and benefits, and staff and resources (see Squire 2017, 2007). Bowen and Greene (2014) link higher levels of legislative professionalism to a higher quality of representation, citizen attitudes and satisfaction with the state legislature, and adherence to institutional rules. Owings and Borck (2000) report a positive association between legislative professionalism and per capita government spending. In a similar vein, Williamson, Morris, and Fisk (2021) find that states with higher levels of legislative professionalism tend to invest more than the minimum amounts of state resources into the state's CWSRF program.

Structural Explanations

The category of structural explanations can be thought of as a set of factors that are generally fixed in time, but that vary across states. In effect, these factors are contextual variables that remain relatively stable (or change slowly) over time. For example, Hoornbeek (2011, 2005) includes a measure of whether a state borders a large body of water; he posits that coastal states are likely to have a greater commitment to nonpoint-source pollution policy than are states that do not have a coastline. Clark and Whitford (2010) include a variable for the land area of a state, and find that states with more land area tend to spend more on environmental programs than do smaller states. Hoornbeek (2011, 2005) includes a variable to represent the percentage of federal lands within a state, but this variable does not reach statistical significance. However, Williamson, Morris, and Fisk (2021) include a measure of state centralization drawn from the work of Stephens (1974) and Bowman and Kearney (2011), and report a negative and statistically significant relationship between centralization and additional state CWSRF spending.

The other element of structural explanations is the EPA region into which a state falls. The regional offices were created during the Nixon presidency, and were intended to provide a connection between federal agencies and state offices. Regional offices were empowered to work directly with states, and to represent the legal authority of the national agency in terms of technical assistance, regulatory enforcement, and other issues. By providing a more geographically proximate source of agency authority, the regional offices were more able to interpret agency policy given the circumstances of states within the region. Clark and Whitford (2010) found significant differences in state environmental spending based on the EPA region in which the state fell. Conversations with retired EPA regional officials[4] suggest that the culture within the regional offices can vary, as can the level of engagement with state agencies and the experience of regional staff. We might conclude from this that regional offices might have an impact on state action in the water quality arena.

Economic Explanations

The variables commonly used to represent economic conditions are, in large measure, variables that represent measures of state wealth. The most common of these measures is per capita income. Clark and Whitford (2010) and Hoornbeek (2005, 2011) use this measure as an indicator of economic well-being. Morris et al. (2021) and Williamson, Morris, and Fisk (2021) substitute median income, which has the advantage of lessening the effect of outliers in each state (see O'Sullivan and Rassel 1995, 330–332). The use of income measures indicates the relative wealth of citizens in each state, but do not speak to the ability or willingness of government to invest in water quality. In addition, the significance of these measures is mixed, with some models indicating a positive (and significant) association, while other research reports no statistical significance for income measures.

Three other examples of water quality research attempt to address this shortcoming by employing alternative measures of economic well-being. Travis and Morris (2003) incorporate a measure of state fiscal comfort index from Tannenwald (1999), which creates an interval-level index that compares an index of the tax capacity of the state with an index indicating the fiscal needs of the state. They report that as the level of fiscal comfort decreases, states become more likely to engage in privatization in the CWSRF program, a relationship significant in all of their models. Mullin and Daley (2018) use a measure of Gross State Product (GSP) as an indicator of state economic health, and they report a statistically significant relationship between GSP and state and local spending on water quality. Finally, although the work does not include an empirical test, Johnson (1995) suggests that a state's ability to pay is an important element of state decision-making in water quality policy.

Sociodemographic Explanations

The inclusion of sociodemographic variables tends to cluster around just two related measures: state population and state population density. State population is the more commonly used measure, but the evidence of the importance of the variable is somewhat mixed. Mullin and Daley (2017), Clark and Whitford (2010), and Hoornbeek (2005) report that state population is a significant variable in their models, while Morris et al. (2021) and Williamson, Morris, and Fisk (2021) find no statistical significance. The second variable, population density, is included by Morris et al. (2021), Fowler and Birdsall (2020), and Williamson, Morris, and Fisk (2021), and shows mixed results.

One potential danger regarding the use of population in models that include measures of federal appropriation is the fact that at least a portion of the federal funding formula for CWSRF capitalization grants includes a factor for state population, so more populous states tend to receive proportionally larger capitalization grants (see Dilger 1986). This has the potential to create multicollinearity issues in regression models. On the other hand, the use of population density makes more intuitive sense, in that the vast bulk of CWSRF resources have been spent for municipal wastewater treatment and control. Centralized wastewater treatment systems are infeasible in rural areas because of the large land area to be covered with relatively few users in that area, while in more densely populated areas the systems can be more compact and serve a larger number of users in a more compact space. Thus, it follows that CWSRF program operations would be more likely in more densely populated states.

Needs Explanations

State needs for wastewater treatment are also a significant element of the federal funding formula as found in the Water Quality Act. Between 1988 and 2012,[5] the EPA collected data from states for wastewater needs in the form of a needs assessment, a process completed every four years. Although there are a number of shortcomings with the methodology and reporting, these needs assessments represent the best data available to determine state needs for water quality.[6] A number of studies (Mullin and Daley 2017; Travis, Morris, and Morris 2004; Morris et al. 2021; Williamson, Morris, and Fisk 2021) have incorporated measures of state need in their models, and the variables typically show statistical significance. For example, Travis, Morris, and Morris (2004) found a strong relationship between state water quality needs and a state's propensity to engage in leveraging. Morris (1997) also noted a significant relationship between needs and the prevalence of private sector actors in CWSRF administrative structures.

Two other measures of state need are found in the extant literature. The first of these, percentage of state waters listed as impaired, is a measure of surface water quality in the state. Separate measures are reported by the EPA for rivers and streams, and for lakes. The greater the percentage of impaired waters, the greater is the need for water quality funds. This approach is employed by Hoornbeek (2005, 2011), who does not find the measure statistically significant in his models. One potential reason for the lack of significance lies in the quality of the data. Conversations with current and retired EPA officials indicate a significant measurement problem with this variable, in that there is no common protocol for measurement and tracking. For example, some states measure every 40 miles of waterway once per year, while other states measure much smaller segments much more frequently.[7] The methodology differs even within states. As an example, Morris et al. (2013) detail the testing protocol on the Nansemond River in southeastern Virginia, a process conducted weekly by members of a local environmental collaborative. The water samples are taken in 2-mile increments and sent to the state laboratory for testing. On the other hand, on a more impaired waterway nearby (the Elizabeth River), testing is done by a combination of state, university, and private sector entities, at random locations along the river and at random times. On the nearby Lynnhaven River, the Department of Health collects water samples weekly, but more frequently following periods of heavy rain. These samples are also targeted for the presence of coliform bacteria and *E. coli*, which can threaten the safety of shellfish farmed in the river; much less attention is paid to other forms of pollutants. The result is a patchwork of protocols on three watersheds within a 10-mile radius. Johnson (1995) recommends the inclusion of a measure of environmental need in water quality studies but does not offer a specific measure to meet that requirement.

Fowler and Birdsall (2020) include a measure of the total pounds of water-based toxic releases per state per year as the dependent variable in their study of primacy under the Clean Water Act. Drawn from EPA's Toxic Release Inventory (TRI), Fowler and Birdsall note several problems with TRI data, including inconsistent reporting, a lack of specificity about the chemicals in the TRI and their relationship to the NPDES permitting system, and the voluntary nature of data reporting. It is also the case that the TRI only captures toxic chemical releases; it does not track nitrogen or bacteria, two of the most common elements of municipal effluent. Although they argue that these data are the best approximation of ambient water quality available, the problems associated with the data make its use as a state-level outcome of water quality efforts problematic.

Environmental Commitment

A common explanation for state behavior in state environmental policy is the level of commitment in a state for environmental policy. These measures include broad measures of state commitment, such as that offered by Lester (1989),[8] as employed by Travis, Morris, and Morris (2007). This measure is reported to be a significant predictor of both a state's decision to leverage CWSRF funds, as well as the aggressiveness of the leveraging scheme. Other measures employed in the literature focus more specifically on state commitment to water quality. Bunch (2008) measures the size of state-run and state-funded programs to water quality, while Travis, Morris, and Morris (2007) add a measure of state commitment to clean water as developed by Lester (1989). Clark and Whitford (2010) include a variable that measures a state's assumption of primacy in water quality, but the variable fails to reach statistical significance. This is likely due to a combination of specification error and the fact that, for the time period of the study, most states had assumed primacy, so there was little variation.

The scholarly literature on primacy is not well-developed. Two early studies by Patricia Crotty (1987, 1988) described the concept of primacy and examined the role of federal agencies in developing primacy decisions in states. A number of studies (see Ringquist 1993; Hunter and Waterman 1996; Travis, Morris, and Morris 2004) employ primacy as an indicator of state commitment to environmental protection. These studies generally report significance of the variable, leading many to conclude that assumption of primacy is an indicator of a state's "greenness" (Woods 2005). However, Woods' (2005) study of primacy suggests that assumption of primacy is only orthogonally related to other factors generally thought to be indicative of commitment to environmental protection; more interestingly, Woods reports a negative correlation between assumption of primacy and policy innovation. However, primacy still plays an important role in water quality, as it determines a state's role in monitoring and enforcement of water quality standards.

The last two measures of environmental commitment noted were included by Hoornbeek (2011). In one portion of his study, Hoornbeek creates a measure of a state's level of policy activism in nonpoint-source pollution and then employs the variable as a dependent variable; his model explains about 71 percent of the total variance in his measure. While this measure is useful to understand commitment to nonpoint-source pollution, the percentage of CWSRF resources allocated by states to nonpoint-source needs remains very small. Finally, Hoornbeek (2011) develops a measure of NPDES authorizations within a state. He uses that measure as a starting point to create a measure of the stringency of state-issued NPDES permits that he also uses as a dependent variable. The variance explained in this model is somewhat less robust at about 43 percent. However, the validity of these latter two measures as indicators of state commitment to the environment are worthy of further exploration.

Interest Group Influence

A significant literature examining the effects of interest group influence in US policy making may be found in the literatures in public policy and political science. In the arena of water quality, measures of interest group influence tend to gravitate to two competing groups of interests: manufacturing interests (generally opposed to environmental policies and regulation) and environmental groups (typically supportive of more stringent environmental protections). Surprisingly, however, these variables have not penetrated deeply into the water quality literature. Clark and Whitford (2010) include a measure of GSP in the manufacturing sector, and find the variable a significant predictor of state environmental spending. The measure does not achieve significance when used as predictor of EPA expenditures, which is likely a function of the nature of the funding formulae for most federal grant programs. Hoornbeek (2005) uses measures of environmental group strength in a state and of industry strength to assess a state's commitment to nonpoint-source pollution, and the same variables to assess the stringency of state NPDES permits (Hoornbeek 2011). Industry group strength fails to reach significance, while environmental group strength is statistically significant for nonpoint-source activism but not for permit stringency.

In terms of the CWSRF program, it is likely that interest group influence would be, at best, marginal. The CWSRF program tends to be regarded, even by states, as more of a technical program, and thus tends to "fly below the radar" of environmental and industry interests. The CWSRF program is largely concerned with the distribution of loans to municipalities, and is thus a funding mechanism to transfer funds from state government to local governments. Although the NPDES process is the embodiment of regulation and enforcement in water quality, it is a separate function from the funding mechanism of the CWSRF. Moreover, the NPDES permitting process has existed for many decades, and its terms and requirements are well understood. Interest groups might clash over questions of permit issuance or enforcement, but funding for infrastructure, the core of the CWSRF program, is much less likely to engender interest group conflict. For this reason, it is unlikely that interest group influence will be a significant explanator of loan distribution patterns in the CWSRF program.

Administrative Factors

One of the central assumptions of Reagan federalism was that states were both willing and able to assume responsibility for federal programs such as the CWSRF program. States differed widely in their previous experience with water quality funding programs. For a majority of states, the experience was largely limited to implementation of the state portion of the construction grants program, which allowed for only a limited state role (see Chapter 4). Other states had small state-funded grant programs, and some states (e.g., Massachusetts, Pennsylvania) had state-administered and state-funded loan programs. As a result, some states had sophisticated and experienced administrative structures in place at the point of CWSRF implementation, while other states scrambled to develop the necessary administrative capacity. States also differed in the degree of political support, state funding, organizational resources and, in particular, the availability of trained budgetary and financial expertise.

The extant literature sheds limited light on the importance of these factors. In terms of the CWSRF program specifically, Heilman and Johnson (1991) first reported on the degree to which states were prepared to assume responsibility for the CWSRF program; Morris (1994, 1999a) also examines the issue of initial state sufficiency of resources. Heilman and Johnson (1991) identify and describe the specific categories of resources required for implementation, and Morris (1994, 1999a) finds a link between loan distribution patterns and the initial sufficiency of administrative, political, and financial resources. Johnson (1995) hypothesizes a relationship between management experience in the CWSRF program and program outcomes, but does not provide an empirical test of this hypothesis.

In related research, Travis, Morris, and Morris (2004) and Breaux et al. (2010) employ a measure of institutional capacity developed by Bowman and Kearney (1988), and find a negative relationship between the decision to leverage and the level of institutional capacity in the state. The inverse relationship reported would seem to contradict expectations, in that leveraging requires a much higher level of administrative and financial expertise; we would thus expect that states that choose to leverage would have the administrative and financial resources to engage in leveraging. However, the findings reported suggest that leveraging is driven by other factors (commitment to the environment, political culture, etc.) rather than institutional capacity. States that leverage are also more likely to engage outside consultants in their administrative structures (see Morris 1999a, 1996). The use of private consultants in CWSRF administration is discussed in greater detail in Chapter 7.

Program Structure/State Choice Explanations

The final set of explanations is centered on the choices that states make in the design, operation, and administration of their CWSRF program. While each of the fifty states has adopted a unique structure for their program, there are a number of similarities that may be detected across the states. More importantly, these structural patterns have been linked to differences in program outcomes; these findings make state choices about structure and operation both relevant and interesting.

States vary in terms of the actors involved in the CWSRF program. Heilman and Johnson (1991) noted that states differed significantly in terms of the mix of environmental and financial personnel in the CWSRF program. Some states included advisory boards or political oversight bodies, while other states did not include these elements. Morris (1997, 1994) noted the significant use of private sector actors in some states, and found that loan distribution patterns differed substantially in states that included private actors. In short, this research determined that private actors could be either involved formally (i.e., under contract), or informally (i.e., informal consulting, and lacking a specific role in the administration of the program). Whether the involvement was formal or informal, Morris (1997, 1994) reported that as private sector involvement in the program increased, states were more likely to make loans as would a private bank; that is, ability to repay the loan overwhelmed environmental need. States with higher levels of private sector involvement were also less likely to make loans to small communities or financially at-risk (hardship) communities, two categories of recipients specifically of interest to Congress as reflected in the Water Quality Act.

Another common structural variable found in the literature is the nature of the state agency in which the CWSRF program is primarily housed (Hoornbeek 2011, 2005; Travis and Morris 2003; Breaux et al. 2010; Ringquist 1993). In the early years of the CWSRF program, the most common type of primary agency was a state environmental agency. Several states also housed their water quality programs in state departments of health, and a small number of states placed their CWSRF program in some form of infrastructure finance authority. Another important difference was the number of state agencies involved in CWSRF implementation. As reported by Heilman and Johnson (1991), the majority of states housed their program in single state agency. However, a significant number of states divided program authority and responsibility across two or more agencies, and at least one state reported a structure that involved six state agencies. The early literature in implementation studies speaks to the complexity of multiple agencies in organizational decision-making. Simon (1948) identified decision-making as the most important process in organizations, and Pressman and Wildavsky (1973) demonstrated the degree to which decisions can be altered by having multiple agencies and decision makers involved in program implementation. Montjoy and O'Toole (1984) further refined Pressman and Wildavsky's work, and quantified the degree to which an additional decision point can alter decision outcomes. In short, complexity matters.

Other elements of state choice found in the literature are more program-specific. For example, several studies examine the degree to which the decision to leverage CWSRF funds impacts both structure and program outcomes. Heilman and

Johnson (1991) note a link between leveraging and program complexity, in that states that leverage tend to have more actors in the CWSRF structure spread across a larger number of agencies. Travis, Morris, and Morris (2004) determine that the choice of whether to leverage is only the first of two important policy questions; the second is how much to leverage. Bunch (2008) reports that states can use either a reserve method or a cash-flow method to leverage. In New Jersey, for example, which employs a cash-flow model, applicant communities actually receive two loans: one at a zero percent interest rate from the Department of Environmental Protection, and another from the New Jersey Environmental Infrastructure Trust. In the reserve model, states issue bonds and use the federal capitalization grant and state match as a reserve. The reserve fund provides security for the bonds, and the interest earned from the reserve account subsidizes the interest rate charged to communities.

States also make program decisions regarding interest rates charged, as well as decisions about whether to target CWSRF resources to certain classes of community. Heilman and Johnson (1991) and Bunch (2008) note the wide range of interest rates charged, ranging from zero percent to rates above the prevailing market rate. Bunch (2008, 121) examines rates typically charged in 2001, and reports a range from 0 percent to 4.3 percent, with a national average of 2.4 percent. Heilman and Johnson (1991) note that some states in the early years of the program offered subsidies in the form of negative interest rates, although this practice was uncommon. Morris (1999b) finds that states that do not leverage are more likely to adopt targeting practices to enhance the distribution of CWSRF funds to communities other than those with significant environmental need.

With the passage of the Safe Drinking Water Act (P.L. 93-523) in 1996, Congress created the Drinking Water State Revolving Fund (DWSRF) program. Modeled exactly on the CWSRF, the Safe Drinking Water Act also empowered states to transfer funds from one revolving fund program to another, largely as they saw fit. Mullin and Daley (2018) found that a state's willingness to transfer funds between revolving fund programs was positively associated with a change in combined state and local spending on wastewater infrastructure, but not for combined spending on drinking water infrastructure. Their findings were likely due to the relatively larger cost of wastewater projects as compared to drinking water projects. States must therefore decide whether to transfer funds between the programs, and the direction of the transfer. Mullin and Daley (2018) also reported a significant and positive relationship between combined wastewater spending and wastewater needs.[9]

A Model of State Choice and Resource Distribution

From the explanations discussed in the previous section of this chapter, we can build a model of state choice in the CWSRF program. As depicted in Figure 6.1,

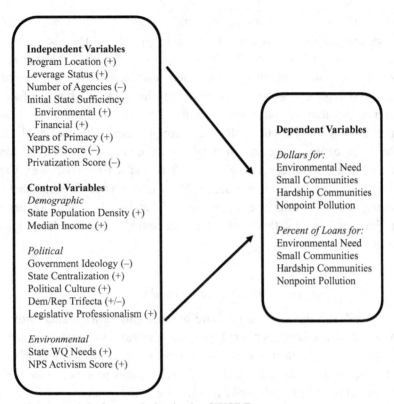

Figure 6.1 A model of state choice in the CWSRF program

the model consists of eight independent variables that represent dimensions of state choice in the CWSRF program. There are also nine control variables included that account for various elements of state context, as described in the extant literature. The data for the analyses cover the time period from the CWSRF program's inception in FY 1988 through FY 2016, the most recent year of stable data.[10] This model will be tested against a series of dependent variables measuring different distribution patterns in the CWSRF program. Appendix A lists the variable descriptions and data sources for each variable.

It should be noted that a significant number of variables used in previous studies were captured in cross-sectional studies, or studies that used a small range of years for analysis. There are thus a number of potentially interesting variables that, while available for a given year (or a small duration of years), are simply not available over a thirty-year time frame. Some of these variables include measures for environmental group membership, industry group strength, and consistent measures of a state's commitment to the environment. To this end, we are limited to variables for which data are available over the span of the study, or that are at least stable over the time period.

In addition, discussions with current and former EPA officials suggest that differences can exist between federal regional offices, and that these differences can determine, in part, state actions within regions. An EPA regional staff that places more emphasis on certain program elements, for instance, might lead states to make different allocation decisions, or may influence state choices of structure. While it makes intuitive sense that regional differences exist, there is nothing inherent in the regional structure, other than simple geography, that might account for differences between regions in a fifty-state comparative study. Moreover, trying to capture differences in regional staff across ten regions, over a span of thirty years, is a significant undertaking. Even if this were feasible, there is no *prima facie* case to assume one region is likely to lead to better outcomes than another. For these reasons, this model does not attempt to treat federal region as a variable of interest.

Dependent Variables

For our dependent variables, we measure the amounts of assistance provided by states to four specific categories of applicants/uses, by year. As noted elsewhere in this volume, the ideal dependent variable would be some measure of change in the impaired waters in each state. Such an analysis would allow us to determine the degree to which state choices ultimately make a difference in the quality of state waters (and, by extension, the nation's waters). However, there are no reliable measures of impaired waters that can be fairly compared across states (see Chapter 1). Given this limitation, we focus our attention on the degree to which states allocate CWSRF resources to the classes of community and applicants that are explicitly mentioned in the WQA and therefore represent the will of the framers of the WQA.

As noted elsewhere in this volume, the Water Quality Act lists several categories of applicants that Congress explicitly intends the CWSRF program to serve; these include communities with significant environmental need, small communities (populations less than 10,000), and financially at-risk communities (also referred to as "hardship" communities). Through the National Information Management System (NIMS) for the CWSRF program, states report annual CWSRF activity to the EPA for inclusion in the NIMS. The reports capture both the number of loans made in each category, as well as the total dollar value of CWSRF assistance given to each category of recipients. This study employs a series of measures of annual CWSRF outputs; we include the percentage of loans made to each of the three categories of recipients, as well as the total dollars awarded to each category of applicant. Because the WQA also places additional emphasis on nonpoint-source pollution, we also include the same two measures (total dollars and percentage of loans) for nonpoint projects.

The use of two forms of measurement (dollars and percentage of loans) is warranted by the rather significant differences in cost between projects. Projects to address significant environmental needs are typically the most expensive, but projects for larger communities are generally more expensive (and larger in scope) than projects for small communities. Likewise, nonpoint projects are generally the least expensive form of project, averaging about $285,014 across all states.[11] Some states have actively sought nonpoint projects and have funded large numbers of these projects, while other states have never supported a nonpoint project with CWSRF resources. By measuring these program outputs by both the number of projects and dollar value we can better assess state commitment for these uses.

Independent Variables

The model includes seven independent variables that collectively represent state choice in the CWSRF program; it is this set of variables that are the focus of the analyses. As detailed in earlier sections of this chapter, all of these decisions represent program design and implementation choices of states, as well as policy choices made by states. These variables include the choice of primary CWSRF implementation agency[12]; the net dollar amount raised through leveraging; the number of agencies involved in CWSRF implementation; the degree to which states allocated sufficient initial resources to implement the CWSRF program; state monetary contribution to the CWSRF in excess of the required state match; the year in which the state achieved primacy in water quality[13]; the number of areas in which a state has achieved primacy under the National Pollution Discharge Elimination System; and the use of private sector actors in the CWSRF structure.[14] Many of these variables are also examined in Chapter 7 in the discussion of initial program design and implementation. Because this set of variables represents the main variables of interest, specific hypotheses are presented for each independent variable.

Program Location

States can choose the institutional "home" of the CWSRF program. The three dominant types of agency are environmental, health, and infrastructure finance. A few states place primary responsibility for the program in a different agency; these are coded as "other" in this study. As a categorical variable, we hypothesize that environmental agencies are more likely to meet environmental need than the other types of agency. Environmental agencies were the dominant agency type in most states for the construction grants program, and are the most commonly charged with the regulatory function in water quality. Therefore, we should see agencies other than environmental agencies to be less likely to fund our four categories of interest.

> H1: States with primary responsibility for the CWSRF program located in an environmental agency are more likely to direct CWSRF resources to environmental need, small communities, hardship communities, and for nonpoint-pollution uses.

Leveraging Dollars

Leveraging provides additional resources with which to make loans to communities. Therefore, states that leverage will have a larger pool of resources, and a greater ability to make loans to applicant communities. The possible exception to this is loans to hardship communities; because of the additional costs imposed by leveraging, states that leverage may not be able to offer a competitive interest rate on the loans.

> H2: States that leverage are more likely to direct CWSRF resources to communities with significant environmental need, small communities, and for nonpoint-pollution needs.

> H2a: States that leverage are less likely to direct CWSRF resources to hardship communities.

Number of Agencies in the CWSRF Structure

As noted earlier in this volume, some states implement their CWSRF program through a single state agency, while other states involve more agencies. Following Pressman and Wildavsky (1973), we would expect that more agencies with program authority would provide concomitantly more decision points, with a greater chance that decisions can be changed. Because of this, states with complex implementation structures, as measured by the number of agencies in the implementation structure, will be less capable of serving the needs of our applicants of interest.

> H3: States with a larger number of agencies in the CWSRF structure will be less likely to direct CWSRF resources to environmental need, small communities, hardship communities, and for nonpoint-pollution uses.

Initial State Sufficiency

States differ in terms of their initial resources on hand to implement the CWSRF program. States that struggled to identify and utilize their resources in the initial years of the program are less likely to be able to meet the intent of the law, while states that had the initial capacity to implement the program have a clear advantage. Given the relatively small variation in several categories of resource sufficiency, along with the inherent tensions in the CWSRF program between environmental and financial elements, we will focus specifically on sufficiency of resources for environmental and financial personnel. Therefore, states with initial sufficiency are more likely to direct resources to the applicants of interest.

> H4: States who report they are initially sufficient in environmental personnel will be more likely to direct resources to communities with environmental need, small communities, hardship communities, and for nonpoint-pollution uses.

> H4a: States who report they are initially sufficient in financial personnel will be more likely to direct resources to communities with environmental need, small communities, hardship communities, and for nonpoint-pollution uses.

Years of Primacy

States choose whether to assume primacy in water quality. A feature of the 1972 Clean Water Act, primacy allows states to assume responsibility for the permitting and monitoring functions in the water quality arena. The variable employed here represents the year in which a state first assumed primacy. The first states to achieve primacy did so in 1973. In a sense, it can be thought of as a measure of both a state's commitment to clean water, as well as a willingness of the state to assume responsibility for water quality in regulatory terms.[15] Because this variable captures the number of years since the state assumed primacy, the relationship in the regression output is expected to be positive.

> H5: States with more experience with state primacy in water quality are more likely to direct resources to communities with environmental need, small communities, hardship communities, and for nonpoint-pollution uses.

NPDES Score

Developed by Hoornbeek (2011), this variable captures the number of areas in which a state has primary regulatory authority in water quality. The variable ranges from 0 (for states who have not assumed primacy) to 5 (for states that have assumed primacy in all available categories). The variable measures a state's willingness to assert state control in the regulatory process; states with greater control of the NPDES process are hypothesized to be more engaged in water quality policy. This suggests that states who assume a greater role in the NPDES process are more likely to assert state control and align their outcomes with the desires of the national government.

> H6: States with a higher NPDES score are more likely to direct resources to communities with environmental need, small communities, and hardship communities.

The NPDES score variable is omitted from the nonpoint-source models. The number of NPDES authorizations held by a state is a measure of a state's commitment to the regulatory framework within the WQA; this framework only addresses point-source pollution. Since nonpoint pollution is not regulated under the WQA but instead left to states, there is no reason to expect a link between CWSRF resources allocated for nonpoint projects and the number of authorizations held by a state for point-source pollution.[16]

Privatization Score

Developed by Morris (1994), this variable measures the degree of both formal and informal private sector activity in the CWSRF program. Morris found that as the level of privatization increased, states were less likely to meet the needs of certain classes of applicant communities. Private actors tend to make allocation decisions more as a private bank would, and ability to repay a loan becomes the primary decision factor. This may be especially true for hardship communities which, by definition, represent a higher category of risk for loan default. Therefore, the expectation is for a negative relationship between a state's privatization score and the dependent variables.

> H7: States with a higher privatization score are less likely to direct resources to communities with environmental need, small communities, hardship communities, and for nonpoint-pollution uses.

Control Variables

The set of control variables employed in these models are divided into three categories: demographic variables, political variables, and environmental variables. The demographic variables include state population density and median income, both of which are expected to generate a positive relationship with the dependent variables. The environmental variables include wastewater needs as reported in the quadrennial Needs Assessment (positively related); the nonpoint models, also include Hoornbeek's (2011) measure of state NPS activism (positively related). As noted earlier in this chapter, there are a range of problems associated with measurements of impaired state waters; although these data are used in some studies, they are not included in the present analyses. In addition, data for impaired waters are only available from roughly 2010 – less than a third of the duration of the period covered by this dataset. Finally, the political control variables include a measure of government ideology; this variable is coded so that higher values indicate a higher level of conservatism; we thus expect the relationship to be negative. Legislative professionalism is expected to indicate a positive relationship. State centralization is expected to indicate a negative relationship. For political culture traditionalistic states serve as the reference group, and the expectation is that both moralistic and individualistic states will produce a positive coefficient. Finally, regarding the variables for split party control of state government, states with a Democratic Trifecta are likely to direct more resources to the applicants of interest, while states with a Republican Trifecta will direct fewer resources. The expected relationships between these variables and the dependent variables are indicated in Figure 6.1 on page 135.

Summary

The extant literature on water quality thus offers a framework to develop a model to examine state choices under the Water Quality Act. While our model includes several variables from previous studies, this model differs from the existing literature in several important ways. First, the model examines program output data over a twenty-nine-year span of time, a much greater amount of time than most existing studies. This allows the assessment of the degree to which the CWSRF program specifically has met the intent of Congress as expressed in the WQA. Second, this model is focused specifically on the effects of state program choices under the CWSRF program. If the WQA is to live up to its promise as an expression of Reagan federalism, then we would expect to find relatively consistent patterns between states. In other words, states may make different choices, but those choices should fulfill the intent of Congress in terms of the distribution of federal resources. Finally, this analysis includes all four of the categories of uses prioritized by Congress in the original legislation: communities with significant environmental need, small communities, financially at-risk communities, and nonpoint-source pollution uses.[17]

The model developed in this chapter provides a framework through which we may examine state choice in the CWSRF program. To better determine the relevance of these variables to state choice, we will examine these variables in three ways. First, Chapter 7 examines initial implementation of state CWSRF programs through a series of descriptive tables. The focus of this analysis is state structural choice, and the analysis relies primarily on data on initial implementation collected between 1989 and 1996. The analyses also compare initial implementation to state program structure in 2020. Chapter 8 presents case studies of four states to illustrate the factors that influenced initial state design, operation, and implementation choices, but also how these choices resulted in their unique loan distribution patterns. Finally, Chapter 9 presents an in-depth analysis of the loan distribution patterns in the states using a series of multivariate regression models. The purpose of this analysis is not only to detect differences in state distribution patterns, but also to explain those differences by employing the different explanations discussed in this chapter.

7

Initial State Implementation of the Revolving Loan Fund Model

With the passage of the Water Quality Act (WQA) in 1987, Congress created a framework for an increased state role in water quality funding. The Clean Water State Revolving Fund (CWSRF) program was designed at the outset to enhance state control and choice regarding water quality funding decisions. The move away from a categorical grant was designed to remove federal control over specific funding decisions and return this authority to states. This trend to return policy authority to state and local governments was fueled by the adoption of policy choices intended to devolve authority from the federal government to state governments. Such policies typically assume that subnational governments are not only willing but also able to assume primary responsibility for the design and implementation of complex national policy initiatives.

This chapter examines the initial implementation of the CWSRF program from a fifty-state perspective. Whereas the following chapter in this volume will examine specific states through the lens of a case study format, the analyses in this chapter are derived from fifty-state surveys of CWSRF state coordinators. The goal is to detect overall trends across the states as each state moved to design and implement their CWSRF program, and to apply for the initial capitalization grant. By focusing on this initial implementation period, we can test the efficacy of the underlying assumptions of Reagan federalism (and thus the logic of the CWSRF itself). The initial period of implementation is important because, although some states have tinkered with the initial implementation structure, the choices made by states in this early period largely determined the state structures as seen today. Likewise, the structures reflected not only the philosophical choices made by states, but they have, in large part, determined the program outputs during the life of the program. A focus on initial implementation can help explain policy outcomes over the life of the CWSRF program.

This chapter consists of three parts. The first part examines the basic structure of CWSRF programs, and the underlying drivers and goals of program design. It also assesses the role of both Environmental Protection Agency (EPA) headquarters and the ten regional EPA offices, and their influence in state design and implementation. The second part focuses on the assumption of initial state sufficiency to assume responsibility for the CWSRF program, and tests assumptions regarding managerial and technical expertise, financial resources, political support, and a willingness to accept the CWSRF program. The final section examines the role of the private sector in the CWSRF program. These analyses include the range of ways in which the private sector was included in state programs, and some of the state conditions that led to a role for the private sector.

Much of the data for these analyses are drawn from two surveys of state CWSRF coordinators conducted in the early 1990s. The first survey, conducted in 1990, was administered under the auspices of a grant from the US Geological Survey to examine initial implementation of the CWSRF program.[1] The survey was long and far-ranging, and included a number of questions about program philosophy, design, operation, structure, administration, and early program outputs.[2] Some of these data were reported in Heilman and Johnson (1991). The survey drew a total of forty-six responses; of the four states that did not respond, two states had not yet received their first capitalization grant. A similar survey conducted by Morris (1994) repeated several of the questions from the 1990 survey, and picked up two of the states that had not responded to the initial survey. The results of that survey are reported, in various forms, in Morris (1994, 1997, 1999a, 1999b).[3] Finally, a survey of CWSRF coordinators was conducted in fall 2020. However, the response rate for this recent survey was a relatively low 44 percent. This chapter reports some results from that survey, but the low response rates for some questions makes comparison across time somewhat problematic.

General Structure, Management, and Administration of the CWSRF Program

The CWSRF program required states to prepare both legal and administrative structures in order to accept a federal capitalization grant. The legal structure required states to enact enabling legislation to create the state program and empower agencies within the state to assume authority to implement the state program. The legislation could either be very general, such as the bill passed in Georgia, or highly detailed, such as the bill in California. In each instance, the legislation laid out the basic structure of the program, and provided the legal authority necessary to implement the program. For states that wished to implement leveraging, this authority included the ability for an entity of the state to issue bonds to support the leveraged scheme. Some states adopted legislation very quickly; Georgia had enacted its legislation nearly a year before final passage of

Table 7.1 *Factors that influenced CWSRF design*

Factor	No. (%) States		
	Major	Moderate/Minor	None
Environmental needs	33 (77)	10 (23)	0 (0)
Federal desire to shift program to states	29 (67)	10 (23)	3 (7)
Build state program capacity	18 (42)	20 (47)	5 (12)
Increase state discretion	13 (30)	25 (58)	5 (12)
Build state financial capacity	31 (72)	12 (28)	0 (0)
Eliminate federal requirements	13 (30)	23 (53)	4 (9)
Address small communities	15 (36)	22 (52)	4 (10)
Personnel	19 (44)	19 (44)	5 (12)
Address non-point	3 (7)	23 (54)	16 (37)
Secure funding	20 (47)	17 (40)	6 (14)

the WQA, and Connecticut and Florida were not far behind.[4] Other states delayed; New York did not pass its enabling legislation until 1989, and Arizona, Montana, Wisconsin, and Wyoming did not pass a bill until 1990.

Upon passage of the enabling legislation, state agencies began the process of designing and implementing the state program. The enabling legislation in some states specified a program structure, while in other states the legislation provided broad authorities to state agencies, and left it to the agencies to develop the specifics of the structure. Across this range, however, there were still many decisions regarding implementation that needed to be considered, developed, and undertaken.

The 1990 survey of state CWSRF coordinators asked respondents about the factors that influenced the design of their CWSRF programs (Table 7.1). The most common factor was a desire to meet environmental needs (77 percent), followed closely by a desire to build state financial capacity (72 percent) and the federal desire to shift programs to states (67 percent). While the latter two categories speak to the question of state capacity, a fundamental assumption of Reagan federalism, it is notable that other elements related to Reagan federalism – build state program capacity, increase state discretion, and elimination of federal requirements – were somewhat less important factors. Also notable is the relative unimportance of the desire to serve small communities and to address nonpoint-pollution needs, two statutory categories of CWSRF needs.

Agency Structure

One of the most important structural decisions concerned the lead agency in the program. Since the early 1970s state water quality programs were generally located in state environmental agencies. The creation of the EPA at the federal level encouraged states to adopt similar agencies at the state level, and many states did so. Prior to this time, many states located their water quality programs in the state health department, and several states kept their water quality program in a state health agency even after the creation of the EPA. With the advent of the WQA, most states chose to keep their existing water quality program in the same agency. However, the WQA also brought a significant requirement for financial expertise. As a result, some states relocated their water quality program into a different agency; most states that relocated did so by placing primary responsibility for the CWSRF program in a financial agency. At the point of initial program implementation, the majority of states (78 percent) located their CWSRF program in an environmental agency.[5] This observation is not surprising, in that the construction grants program that preceded the CWSRF was seen by states as an environmental program. An additional ten states located the CWSRF program in a state health agency; in those states, the water quality program was previously located in the health agency.[6] No programs in 1990 were located in financial agencies, and only one program was located in an agency coded as "other." By 2019, however, a number of states had moved responsibility to different agencies. While the bulk of states (70 percent) still housed the CWSRF program in an environmental agency, only four states housed the program in a state health agency. The more significant change was the growth in the use of financial agencies as the institutional "home" for the CWSRF program; by 2019, nine states employed financial agencies as the lead agency in CWSRF implementation. This is indicative of a growing realization on the part of states that the financial component of the CWSRF model required expertise beyond that typically found in environmental and health agencies.

Because of the need for both environmental and financial expertise, a number of states divided program authority across multiple agencies. Only nine states initially placed the program within a single agency; by 1994 this had fallen to just two states. The majority of states split program authority across two or three agencies. Of note was the trend over the first several years of the program to involve more agencies in the program. In 1990 only three states split the program across four or more agencies; by 1994 more than 40 percent of states involved more than four agencies, and eight states had six or more entities involved in program implementation. As Pressman and Wildavsky (1973) and Montjoy and O'Toole (1984) pointed out, the number of decision points in an implementation structure exponentially complicated the decision processes. In the case of the CWSRF program, the growing complexity reflected the tension between the financial and environmental components of the program.

The tension between the environmental and financial components of the program is evident in Table 7.2. Although the CWSRF is ostensibly an environmental program, 85 percent of respondents rated the financial emphasis in the CWSRF as "moderate" or "major." By comparison, only one respondent rated the financial emphasis in the construction grants program as "major" or "moderate." Compared to the CWSRF's predecessor program, the financial component is clearly a significant change in the water quality policy arena. The results from the 2020 survey largely reflect the results from the 1990 survey.

Another important element to consider in terms of program structure was the presence of state-run/funded alternatives to the CWSRF. The size and scope of state alternatives varied greatly, as did the nature of the state programs (Table 7.3). Slightly over half of respondents operated state loan programs outside the CWSRF, and nine of the states operated revolving loan funds. States with experience in the administration of CWSRF-like programs (straight loan programs or revolving loan programs) had existing structures in place from which to build CWSRF implementation structures. Of interest is the growth in state-funded alternatives from 1990 to 2020. Although the response rate is lower in 2020 than in 1990, both grant programs and state-funded revolving loan funds have become more widely used by states. The growth in dollars in state-funded programs, as reported in the National Information Management System (NIMS) supports this finding.

The Role of EPA in Program Design and Implementation

The Office of Water and the Wastewater Branch in EPA began to develop guidance for states even before the final legislation was enacted. The same working group that had developed the revolving fund proposal in 1984 had also made clear that the new program would place additional demands on states (EPA 1984). In preparation for this, EPA released its draft initial guidance for states in September 1987, the final guidance document in January 1988 (EPA 1988a), and a complete program management manual in December 1988 (EPA 1988b). These documents were accompanied by a series of workshops held both in Washington, DC, and several other cities around the country. These training workshops were designed to provide information to states about how to design, implement, and administer the new water quality program. The workshops focused primarily on technical issues such as the capitalization grant process, the Letter of Credit (LOC), and state reporting requirements. However, the workshops also provided information on both environmental and financial aspects of the new program, with a specific emphasis on leveraging.[7] When asked about the emphasis on the

Table 7.2 *Level of emphasis in financial and environmental elements, CWSRF and Construction Grants*

Level of Emphasis	Environmental Elements				Financial Elements		
	CWSRF91 % (N)	CWSRF20 % (N)	Construction Grants % (N)		CWSRF91 % (N)	CWSRF20 % (N)	Construction Grants % (N)
Major	79.1 (34)	78.3 (18)	100 (44)		64.3 (27)	78.3 (18)	0 (0)
Moderate	18.6 (8)	21.7 (5)	0 (0)		21.4 (9)	17.4 (4)	1 (2)
Minor	2.3 (1)	0 (0)	0 (0)		14.3 (6)	4.4 (1)	43 (98)

Table 7.3 *State alternatives to the CWSRF program*

Program Type	% Yes 1990	% Yes 2020
Grants	71	83
Loans	53	57
Bond pool	31	30
Revolving fund	26	40
Other	21	15

Table 7.4 *US and regional EPA CWSRF program emphasis*

	Regional EPA Offices		EPA Headquarters	
	No. (%) States			
Emphasis	Major/Moderate	Minor	Major/Moderate	Minor
Environment	41 (69)	11 (28)	12 (31)	18 (46)
Finance	31 (80)	7 (18)	25 (64)	9 (23)

environmental and financial aspects of the program by EPA (Table 7.4), the results again highlight the tensions between the environmental and financial elements of the CWSRF model. The data show that both the regional and headquarters' offices emphasized the financial aspects of the program more than the environmental aspects. This is particularly true for the ten regional offices; this role is not quite so strong for EPA headquarters, but the pattern is similar. Surprisingly, almost half of respondents rated the environmental emphasis of EPA headquarters as minor. As the federal government's primary environmental agent, EPA is regarded by state CWSRF coordinators as more concerned with financial issues rather than water quality issues. Results from the 2020 survey are almost identical to the findings reported in the 1990 survey.

In addition to the regional meetings and written guidance, EPA regional offices and headquarters staff were in regular contact with states through the design and implementation phases. Respondents to the 1990 survey were asked about the roles played by both EPA headquarters and the EPA regional offices in the establishment and operation of the CWSRF program in the state (Table 7.5). For the purposes of this analysis, a directive role was defined as the degree to which EPA promoted a set of options and initiatives. The reactive role was defined as EPA's responsiveness to state initiatives. A facilitative role measured the degree to which EPA provided technical assistance, and the regulatory role was expressed in terms of policy compliance and regulatory actions taken by EPA. The table indicates that the EPA

Table 7.5 *Role of US and regional EPA in state CWSRF establishment*

Establishment of CWSRF

No. (%*) States

Role	Regional EPA		EPA Headquarters	
	Major/Moderate	Minor	Major/Moderate	Minor
Directive	21 (51)	19 (46)	11 (27)	24 (59)
Reactive	34 (85)	5 (13)	24 (60)	10 (25)
Facilitative	21 (54)	17 (44)	9 (23)	23 (59)
Regulatory	30 (80)	8 (21)	16 (42)	12 (32)

Operation of CWSRF

No. (%*) States (1990)

Role	Regional EPA		EPA Headquarters	
	Major/Moderate	Minor	Major/Moderate	Minor
Directive	15 (38)	24 (60)	6 (15)	27 (68)
Reactive	29 (74)	9 (23)	14 (36)	18 (46)
Facilitative	19 (50)	16 (42)	9 (24)	19 (50)
Regulatory	34 (92)	3 (8)	16 (43)	11 (30)

No. (%*) States (2020)

Role	Regional EPA		EPA Headquarters	
	Major/Moderate	Minor	Major/Moderate	Minor
Directive	7 (30)	15 (65)	9 (39)	12 (52)
Reactive	16 (70)	5 (22)	6 (26)	14 (61)
Facilitative	15 (65)	7 (30)	7 (30)	14 (61)
Regulatory	19 (83)	2 (9)	13 (57)	3 (13)

*Percentages do not equal 100 because this table omits respondents who answered "Other/Don't Know."

regional offices were more directly involved with states than was EPA headquarters in the establishment phase of the CWSRF. However, the role was more often described as reactive and regulatory than directive or facilitative; the same pattern exists for regional offices. However, the role played by both the regional offices and EPA headquarters generally declined after the establishment of the program as states entered the operational phase, with the exception of the directive role of EPA headquarters; CWSRF coordinators reported an increase in this role in 2020. This is largely consistent with the intent of WQA, and of Reagan federalism, in that the CWSRF program has reduced the role of federal government actors in the administration of water quality programs. However, the enforcement and regulatory power of EPA is still clearly felt by states.

Speed of Implementation

As noted earlier in this chapter, states moved at different speeds when implementing their CWSRF programs. Some state timetables were driven by state legislative calendars, while others were driven by the interest of state officials to either move the program forward, or to delay implementation. Slightly over half of respondents reported they moved rapidly with implementation, while about a quarter of respondents reported moving slowly, and another quarter of respondents reported moving neither rapidly nor slowly. When asked about the factors that influenced implementation speed, stark differences between states that moved rapidly and other states become evident (Table 7.6). States that moved quickly were driven by a need for funds, a desire to assume control of their water quality program, a belief they were an innovative state, and a desire to shift away from a grants program and reduce the requirements that accompanied the Title II program. On the other hand, states that did not move quickly with implementation preferred a grant program, were concerned with limited state resources, and needed funds. While none of the states that implemented rapidly wanted to wait to evaluate the new program, half of those who implemented slowly played, in effect, a waiting game.

The differences in implementation speed highlight the inherent tensions in the CWSRF program, as well as the inherent tensions of federalism. More than half of all states wished to retain a grants program, and the pending loss of a grant program clearly influenced some states to delay implementation of the revolving loan program.[8] Moreover, the differences in implementation speed support the conclusion that not all states were adequately prepared to assume responsibility of the CWSRF program. States that implemented more slowly had less interest in assuming control of the program, more concerned about the availability of state resources, and saw themselves as less innovative. Heilman and Johnson (1991) and

Table 7.6 *Factors influencing speed of CWSRF implementation*

| | All States N (%) | | Implementation Speed of State — N (%) States | | | |
| | | | Rapidly | | Slowly/Neither | |
Factor	Major/Moderate	Minor/Not Important	Major/Moderate	Minor/Not Important	Major/Moderate	Minor/Not Important
Desire to retain grant program	24 (57)	18 (43)	7 (32)	15 (68)	14 (82)	3 (18)
Needed funds	29 (67)	14 (32)	16 (60)	11 (40)	11 (61)	7 (39)
Wait to evaluate CWSRF	9 (21)	34 (79)	0 (0)	23 (100)	9 (50)	9 (50)
Concern about solvency of CWSRF	17 (40)	26 (61)	9 (41)	13 (59)	8 (44)	10 (56)
Wanted to use CWSRF as complementary to state programs	19 (44)	24 (56)	13 (59)	9 (41)	5 (28)	13 (72)
Wanted to assume control of program	24 (56)	19 (44)	17 (77)	5 (23)	6 (33)	12 (67)
State had limited resources to assume program	19 (44)	24 (56)	6 (27)	16 (73)	11 (61)	7 (39)
Innovative state	26 (61)	17 (39)	18 (82)	4 (18)	6 (33)	12 (67)
Wanted to shift from grant program	27 (63)	16 (37)	20 (91)	2 (9)	6 (33)	12 (67)
Wanted to reduce Title II requirements	32 (74)	11 (25)	18 (82)	4 (18)	11 (61)	7 (39)
Believed federal government provide other options	7 (16)	36 (84)	3 (14)	19 (86)	3 (17)	15 (83)

Morris (1994) noted that states that implemented more slowly made fewer loans in the early years of the program, were less likely to leverage, and were less likely to meet the categories of community Congress desired to be served by the CWSRF program.

A New Game: Initial State Sufficiency

As discussed in previous chapters, a fundamental assumption of the CWSRF program was that states would have the technical, managerial, administrative, political, and financial resources to design, implement, and administer a complex financial program. While the EPA provided a great deal of initial technical assistance,[9] implementation still rested entirely with states. The degree to which the CWSRF program is able to meet water quality needs thus rests on the decisions made by states, and the ability of state administrators to design and implement a viable state program. In this regard, the CWSRF is a clear test of the underlying assumption of Reagan federalism that states possess the knowledge, resources, and willingness to implement the program.

There are several potential categories of resource that are brought into focus in this analysis (see Table 7.7). First, do states possess the technical and managerial resources necessary to administer the program? As noted elsewhere in this volume, the CWSRF model places a great deal of emphasis on the financial components of the program, particularly for states that choose to leverage their programs. While such financial expertise is generally available in the private sector, it is somewhat less commonly found in state government, and even more rarely in traditional

Table 7.7 *Categories of initial resources for CWSRF implementation*

Category	Definition
Technical and managerial resources	Experienced personnel to address the technical requirements of projects, address environmental requirements and concerns, and monitor construction of facilities; adequate financial expertise in state government.
Financial resources	Sufficient funding to administer the program. This includes not only state funds to fund administrative costs (above and beyond the 4 percent available from the federal capitalization grants), but funding for the required 20 percent state match.
Political support	Political support in the state legislature, the governor's office, and among interest groups and citizens; delegation of adequate program authority (statutory and regulatory).
Willingness to accept	Belief that the CWSRF was not a "stopgap" policy; perceived need among communities in the state for program assistance.

environmental agencies. These state agencies were generally created around the time of the creation of the EPA; nearly all of the environmental programs in place in these early years were categorical grant programs that did not require significant financial expertise on the part of the state agencies. The passage of the WQA created a new imperative for state governments in this regard; how states met this requirement is a fundamental component of the story of the CWSRF program.

Second, and relatedly, are the state financial resources available for water quality. There are two elements in this category. First, because states are now required to assume primary responsibility for most administrative tasks related to water quality (including the production of required program documents such as the Intended Use Plan and the annual report), state agency staff need to be augmented with additional employees to meet the new workload demands. The WQA allows up to 4 percent of each capitalization grant to be dedicated to offset administrative costs, but most states reported that this new infusion of funds still did not meet the increased administrative costs.[10] Second, each CWSRF capitalization grant requires a 20 percent state match; without the matching funds in place, states cannot receive the Letter of Credit required to withdraw funds from the treasury. Even in relatively small states with more modest capitalization grants under the funding formula, the required match represents a significant new budget outlay by the state legislature. Some states, such as Georgia and Alabama, have found ways to meet the federal requirement outside the traditional mechanism of legislative appropriation. And, because the capitalization grants are insufficient to meet current needs, there is additional pressure to ask legislatures to appropriate additional state funds to meet state water quality needs. While some states operate state-funded grant or loan programs, most states rely almost entirely on federal funding assistance.

Third, political support was needed from the state legislature and the governor's office to receive authority to implement the new program. As detailed in an earlier chapter, there were several steps states needed to complete prior to receiving the initial capitalization grant; one of these was to enact enabling legislation to create the state-level program and provide administrative authority to the various state actors involved in implementation. Some states did this very early; indeed, Georgia had passed its enabling legislation nearly a year before the final passage of the WQA itself.[11] Political support was also important in that it helped persuade legislatures to appropriate required matching funds, or allocate additional state funds. Finally, political support was necessary to provide the authority necessary to issue bonds; if a state wished to engage in leveraging, the agency had to have access to a viable bond authority.

Finally, an important element was the willingness of administrative agencies in the state to accept the new program. Data reported by Heilman and Johnson (1991) suggested that several states lagged in their design and implementation because state administrators were convinced that the CWSRF program was a temporary aberration, and that a return to the construction grants program was imminent.[12] Other states lagged back because they did not believe there was a demand in their state for loan assistance. Taken as a whole, this variation in state "buy-in" to the program can tell us much about early state implementation. Likewise, early implementation decisions help determine program outcomes well into the future.

When asked about the sufficiency of initial resources across a set of categories – political, organizational, budgetary, and financial expertise[13] – most states (twenty-two of thirty-seven respondents) reported they had sufficient resources across all four categories, while fifteen states reported they were initially insufficient in at least one category, with the most common insufficiency reported as budgetary resources (twelve respondents).[14] One possible explanation for differences in initial sufficiency is the existence of non-CWSRF state-funded and state-administered water programs. A total of twenty-two states reported they had a state loan, revolving loan, or bond bank program in place prior to the CWSRF program, while twelve respondents reported having no state program.[15] When we compare the existence of state programs with states reporting initial sufficiency (Table 7.8), we see a weak relationship between the variables. The relationship fails to reach statistical significance ($x^2 = 0.3827$; $p = 0.536$); a correlation between the variables is also statistically insignificant ($r = -0.1017$), and negative, suggesting that state programs draw resources away from CWSRF resources (and, perhaps, vice versa). However, the relatively low number of respondents to these questions limits the ability to reach definitive conclusions regarding this explanation.

States were asked about the level of emphasis placed on the financial elements of the CWSRF program. Twenty-nine states reported that they placed "major

Table 7.8 *Cross-tabulation of state-funded CWSRF alternatives by initial state sufficiency*

State-funded loan programs	Initial State Sufficiency		Total
	Not Initially Sufficient (N; Row %; Col %)	Initially Sufficient (N; Row %; Col %)	
No state programs	4 (33.33) (26.67)	8 (66.67) (36.36)	12 (100) (32.43)
State alternatives	11 (44.0) (73.33)	14 (56) (63.64)	25 (100) (67.57)
Total	15 (40.54) (100)	22 (59.46) (100)	37 (100) (100)
	Pearson chi^2(1) = 0.3827 Pr = 0.536		

emphasis" on the financial elements of CWSRF, while nine states reported they placed "moderate emphasis" on the financial elements. In spite of the seemingly strong emphasis on the financial side, states still view, or at least report, the CWSRF program in environmental terms. When asked about the emphasis placed on the environmental elements of the CWSRF, thirty-four states indicated "major emphasis," while eight states indicated only "moderate emphasis" on the environmental elements of the program. More telling, perhaps, is the finding that six states placed "minor emphasis" on the financial elements, while only one state reported "minor emphasis" on the environmental aspects of the program.

An important issue in the CWSRF program is how states address the need for financial expertise to administer the demanding financial components of the CWSRF program. The most-often reported strategy was to use in-house resources to address the financial requirements of the program (Table 7.9). Many states (twenty-eight) also placed part of the CWSRF administrative structure in another state agency (typically a state treasurer, state auditor, or similar agency). While some states (seventeen) employed new personnel to address the financial component of CWSRF, twenty states employed outside consultants to address the financial requirements of the program. Many states used a combination of tactics, thus ensuring that programmatic authority in the CWSRF program was dispersed among a variety of actors, whether in the public sector or private sector. Not surprisingly, many of the states reporting the use of leveraging also report the use of outside consultants in the CWSRF program. States that use outside consultants also tend to place a high degree of emphasis on the financial elements of the program, and tend to report a major overall role for the private sector in the CWSRF. States have generally been consistent over time in their sources of financial expertise; the results from the 2020 survey are very similar to those from earlier surveys. Indeed, a case-by-case analysis of states that responded to both surveys indicates little variation across the thirty-year span between the surveys.

Table 7.10 suggests that while both leveraged and non-leveraged states are equally likely to both use in-house resources and to place part of the CWSRF

Table 7.9 *Source of financial expertise for CWSRF*

Source	Percent* (N) Yes 1990	Percent* (N) Yes 2020
Used in-house resources	70.5 (31)	69.6 (16)
Employed new personnel	38.6 (17)	26.1 (6)
Used outside consultants	45.5 (20)	52.1 (12)
Placed financial function in another state agency	63.6 (28)	39.1 (9)
Other	4.5 (2)	4.3 (1)

*States may appear in more than one category.

Table 7.10 *Source of financial expertise for CWSRF, by leveraging status*

	All States	Leveraged States	Non-Leveraged States
Used in-house resources	70.5 (31)	69.2 (9)	71 (22)
Employed new personnel	38.6 (17)	53.8 (7)	32.3 (10)
Used outside consultants	45.5 (20)	69.2 (9)	35.5 (11)
Placed financial function in other state agency	63.6 (28)	69.2 (9)	61.3 (19)
Other	4.5 (2)	7.7 (1)	3.2 (1)
Percents given; No of states in parentheses			

structure in another agency to meet the financial requirements of the program, leveraged states are nearly twice as likely to employ outside consultants to help manage the CWSRF financial structure.

In sum, the assumption of initial state sufficiency across all states is clearly unsupported by the analyses. While some states clearly had the resources required to assume responsibility for the CWSRF program, most states lacked at least some important element, or struggled to develop their existing resources to assume program responsibility. States were generally prepared to assume an environmental role, but this is unsurprising given that the environmental components of the CWSRF program did not differ substantially than those in the Title II program: States still developed priority lists,[16] worked with applicants to develop projects and provide engineering expertise, and continued to monitor compliance. The biggest issue was the financial component, in that most traditional state environmental agencies did not have the expertise to assume responsibility for a complex financial structure. Some states foresaw this program by creating implementation structures that included state entities with financial expertise in the CWSRF structure, while other states turned to the private sector for that expertise. Political support also varied, as evidenced by the time it took some states to enact enabling legislation and secure state matching funds from the state budget. While five states had enabling legislation in place by the end of 1987, twenty-five states enacted legislation between 1989 and 1991. This meant that eight states had received their initial capitalization grant by FY 1988, but another eight states did not receive their initial grant until FY 1991.[17]

Privatization in the CWSRF Program

Much of the literature on privatization revolves around the affirmation of the privatization model as a strategy to close the gap between increasing demands for public goods and services, and decreasing resources with which to meet those demands. The CWSRF program involves a different kind of privatization. Rather than producing direct and tangible goods and services, privatization in the CWSRF program incorporates the private sector in public policy implementation – the design, implementation, and administration of the SRF program. The nature of the need for these services is fundamentally different than the kinds of services delivered or produced in many privatized arrangements more commonly found in the academic literature.

The tremendous complexity and requirements of the financial structures in the CWSRF model impels all states to address both the financial and environmental objectives of the program and some states to seek financial services from the private sector. The tension in the CWSRF policy instrument between the objectives of environmental quality and financial solvency provides an opportunity for substantial private sector influence in the CWSRF program. Rather than making decisions based solely on the environmental need of the potential recipients, states must now consider the long-term viability of the financial structure used to supply funds for wastewater construction. While states typically have many years' experience in the implementation of federal environmental standards, the financial requirements of the CWSRF program are new to many states. The financial expertise needed to run a complex loan program is traditionally not found in state water quality agencies. As a result, some states have had to look elsewhere to find needed financial expertise. For many states, the most available source of this expertise is the private sector.

As detailed in Morris (1994) there was wide variation in both how states accomplished this, as well as the degree to which states engaged with the private sector in program implementation and administration. While Georgia, for example, did not engage with the private sector at all, the program in Alabama was, for all intents and purposes, run entirely by a private sector consultant. Most states that engaged with private sector actors did so through a series of formal contracts and agreements, although some states engaged private sector actors in a much more informal manner. Morris (1994) analyzes the impact of these choices on the distribution of program resources (loans); the nature and amount of privatization in the CWSRF program was a significant predictor of loan distribution patterns. Moreover, the analyses concluded that privatization, contrary to its depiction in the academic literature, is not a unidimensional concept. Rather, privatization can be described as either formal or informal, depending upon the nature of the relationship between the principal (government) and the agent (private sector). Privatization may intrude into the provision of a service, in addition to the production and/or delivery of a service. When this occurs, the sovereignty of the state may be challenged with resultant negative implications for the allocation of public goods and services.

Respondents to the 1990 survey were asked to evaluate the overall role and influence of the private sector in their CWSRF program. Eighty percent of states replied that the private sector played some role; only 20 percent of states reported that the private sector played "no role" in the CWSRF. Although the difference between those reporting a "major role" and those reporting a "minor role" is small, the expanded role for the private sector in the CWSRF program[18] is striking. If one considers the number of states reporting *any* role for the private sector (thirty-five of forty-four states), the assertion that the design of the CWSRF program provides new opportunities for the private sector appears to receive additional support.

The 1994 survey collected additional detail regarding the scope and role of private sector activity in the CWSRF program. States were asked to indicate the agency or actor who had either formal or informal authority to act across a set of ten decisions in the CWSRF program (see Table 7.11). For the sake of simplicity, the table shown only includes those that indicated a private sector role. The striking result from this question is the wide range of design and administrative activities in the CWSRF program that involve private sector actors. Across the ten items, private sector actors are involved in all ten decisions, either formally or informally. Concomitantly, more states report informal authority exercised by private sector actors across the set of programmatic decisions.[19] This table also illustrates the tensions between the environmental and financial imperatives of the CWSRF program. While private sector actors are commonly involved in the core financial elements of the program, private actors also have a prominent role in the

Table 7.11 *Formal or informal role in CWSRF program by function*

Decision	Privatization Role		
	Formal (No. of States)	Informal (No. of States)	Total
Establish the financial structure of the CWSRF	1	7	8
Set priority list	0	3	3
Set interest rates for loans	3	6	9
Set loan payback provisions	4	7	11
Evaluate loan applicants for environmental need	0	3	3
Evaluate loan applicants for creditworthiness	3	3	6
Determine who gets a loan	2	3	5
Set size of loan	3	5	8
Monitor loans (payback)	3	4	7
Audit loans	10	3	13

environmental decisions – setting the priority list and evaluating applicants for environmental need. These decisions also have real consequences for the distribution of loans from the program (Morris 1994, 1997). When private actors are involved in decisions regarding creditworthiness, setting the size of the loan, or determining who gets a loan, states effectively cede the core function of the public sector to the private sector – the ability to make decisions about the distribution of public resources.

States that engage in leveraging are also more likely to include private sector actors in the CWSRF. A cross-tabulation of leveraging status and the role of private sector actors in the CWSRF (Table 7.12) illustrates this relationship. States that operate a straight loan program are much less likely to include a major role for private actors in the CWSRF program. This serves to further illustrate the administrative complexities that accompany a decision to leverage in the program.

A comparison of the role of the private sector in the early stages of SRF implementation[20] reveals a marked difference in the scope and role of the private sector in SRF between leveraged and non-leveraged programs (Table 7.13). Leveraged states report substantially greater roles for the private sector in all phases of the SRF program, from the design and development of the SRF structure through its administration. Of particular note in this table is the role for the private sector in leveraged states in setting the priority list. While there is a small amount of this activity in non-leveraged states (two non-leveraged states report minor

Table 7.12 *Cross-tabulation of the role of private sector actors in CWSRF, by leveraging status*

		Leveraging (N)	
		Yes	No
Role of the private sector in CWSRF	No/Minor Role	2	24
	Major Role	11	7

Table 7.13 *Mean number of private sector actors in early stages of CWSRF, by leveraging status*

	All States	Leveraged States	Non-leveraged States
Development	1.33	1.79	1.12
Design	1.53	2.33	1.15
Administration	1.16	2	0.77
Develop financial system	1.08	1.82	0.77
Set priority list	0.34	0.82	0.15
Set interest rates	0.74	1.64	0.37

private sector involvement in this activity), leveraged states have clearly allowed the private sector to assume a financial role in a substantially environmental activity. Furthermore, the practice of allowing a role for the private sector in this activity is not limited to a few leveraged states. Forty-six percent of the leveraged states allowed at least some role for the private sector in setting the priority list, compared with 23 percent of the non-leveraged states. All of the non-leveraged states that report private sector activity in setting the priority list describe only a minor role for the private sector, while three of the leveraged states report a substantial role for the private sector.

Summary

The choices made by states in the early years of implementation are key to understanding program operation and outcomes. In these early years of program design and implementation, states made choices that largely determined the function, operation, and emphasis of the CWSRF program for years into the future and, ultimately, determined the patterns of loan distribution and the ability of states to meet their water quality needs. In this sense, the early design and implementation of the CWSRF program in states becomes the first "test" of the underlying assumptions of Reagan federalism. These results also demonstrate that decisions made at the outset of the program can have meaningful and lasting impacts even three decades later, and can thus impact the ability of states to provide their citizens with clean water.

There are several conclusions that may be drawn from this analysis. First, the assumption of state capacity to assume responsibility for a complex funding program was clearly not met in all states. While some states not only possessed the capacity, they also implemented the CWSRF program quickly, and in a manner consistent with the intent of Congress. These states made decisions based on the context of water quality needs and demands for assistance in their states. States that moved ahead quickly with CWSRF implementation were driven by a combination of a desire to shift away from a grant program; wanted to assume state control of the program; saw themselves as an innovative state; and wanted to reduce the federal requirements that accompanied the Title II program. On the other hand, states that moved forward more slowly with implementation were less likely to see themselves as an innovative state; wanted to retain the Title II (construction grants) program; did not have the resources necessary to assume the CWSRF program; and were more likely to lack the political, organizational, financial expertise, and budgetary resources necessary to implement the program.

Second, states differed significantly in the structure of their CWSRF programs. In addition to the wide variation in the nature of the lead agency for the CWSRF program, some states had relatively simple structures while others involved many more agencies. These factors are likely due to a combination of state history[21] and a decision whether to engage in leveraging. Likewise, states approached the need for financial expertise in their CWSRF structure in different ways. While some states were able to draw on existing state financial expertise, other states sought financial expertise in the private sector.

Third, these analyses highlight the tensions between the environmental and financial elements of the CWSRF model. A major element of Reagan federalism was a desire to devolve federal costs to states, and to increase state financial self-sufficiency in the long term. The CWSRF program was designed to accomplish this goal, but it also introduced significant tensions between the need for long-term fund solvency and the need to meet environmental requirements for clean water standards. This tension is evident in terms of the structural choices made by states, as well as the differing emphases of both EPA regional offices and EPA headquarters. The tension is also reflected when we compare the relative emphases between the construction grants program and the CWSRF program. The financial component of the construction grants program was minimal for states, while it is a significant emphasis for states in the CWSRF program. Likewise, while all states reported a major emphasis on the environmental elements of the construction grants program, only about 79 percent of states saw a concomitant level of emphasis on the environmental elements of the CWSRF program. While the construction grants program was clearly an environmental program, the CWSRF program has what we might refer to as a "split personality," and tries to accomplish two often incompatible goals in the same program. While it is beyond the scope of this analysis, it is likely that the program has continued to operate largely due to the continuation of Title VI appropriations for some twenty-six years beyond the WQA's authorization; this has pumped tens of billions of additional dollars into state CWSRF programs, thus dispelling some of the concerns regarding long-term financial viability.

Finally, the decision to engage in leveraging was another consequential decision made by states. States that leveraged were more likely to hire new personnel or to hire outside consultants to address the need for financial expertise, while states that did not leverage were more likely to rely on either in-house expertise or expertise in another state agency to address the financial needs of the program. Leveraged states were much more likely to report that private sector actors played a major role in the CWSRF than were non-leveraged states. More importantly, leveraged states were much more likely to include private sector actors in all phases of CWSRF design and implementation than were non-leveraged states. As Morris (1997) pointed out, this greater level of involvement in program implementation, particularly in terms of setting environmental priority lists, setting interest rates, and deciding who gets a loan, led to significant differences in loan distribution patterns. In this sense, states do exhibit the characteristics of policy laboratories (Dror 1968), and the nature of the CWSRF program gave states the freedom to realize the philosophy of Reagan federalism. Still, the willingness and ability of states to assume responsibility for the program differed substantially from state to state.

8

Implementation "On the Ground"

Four Case Studies

The patterns detected in the fifty-state comparative analyses of initial state implementation of the Clean Water State Revolving Loan Fund (CWSRF) program presented in the preceding chapter provide evidence of the overall national implementation of the Water Quality Act (WQA). In order to build a more complete picture of the effects of this program, we must focus our attention on the state level to discern how these states act within the context of the WQA. Unlike the construction grants program in place under the Clean Water Act, the fate of the CWSRF rests almost entirely on state choices and actions. More importantly, the initial choices made by states become important, since the early design and implementation choices made by states largely determine ongoing program implementation. We thus adjust our focus to examine these state choices. Table 8.1 compares the four case study states across a set of independent variables from the model presented in Chapter 6. Likewise, Table 8.2 compares the four case study states across the four dependent variables (measured in dollars). The relative sizes of the four programs are evident in this table, as is the variation in dollars allocated to the four classes of applicants in this study.

This chapter presents case studies of four states: Alabama, California, Georgia, and New York. These states are included because they represent states arrayed across a variety of continua, including water quality needs, political and financial support, presence of state-funded alternatives for water quality funding, and initial willingness to adopt the new CWSRF program. The point of the case studies is to examine the conditions in place in the states at the time of initial implementation; at the end of the chapter we will compare the states over the course of the thirty years since the passage of the WQA to determine the degree to which these early decisions shaped long-term program outcomes. The information in the case studies is drawn from on-site interviews conducted with state CWSRF program coordinators and staff conducted in 1990–1991[1] and 1994.[2] In addition, state enabling legislation, Intended Use Plans (IUPs), annual reports, and other

Table 8.1 *Comparison of case study states across select independent variables*

Variable	Alabama	California	Georgia	New York
Political culture	Traditionalistic	Moralistic	Traditionalistic	Individualistic
Primacy year	1979	1973	1974	1975
NPDES permits	4	4	4	3
Primary institutional location	Environmental	Environmental	Financial*	Financial*
Privatization factor**	4.2927	−0.57548	−1.23738	−0.038459
Total actors	4	5	4	2
Leveraged	Yes	Yes***	No	Yes
Initial sufficiency – Env.	Yes	Yes	Yes	Yes
Initial sufficiency – Fin.	No	Yes	Yes	Yes

*Both Georgia and New York initially began operating the program in an environmental agency, but switched to a financial agency during the study period.
**Factor residual from Morris (1994). Higher values indicate greater private sector role.
***California has only leveraged in two fiscal years; 2003 and 2016.

Table 8.2 *Cumulative dollars to dependent-variable categories, case study states*

Dependent Variable*	Alabama	California	Georgia	New York
Environmental need	$889,033,465	$7,473,825,917	$1,282,577,356	$12,996,966,438
Small communities	$314,682,928	$696,317,212	$519,289,350	$1,993,360,000
Hardship communities	$0	$218,600,000	$0	$1,004,348,463
Nonpoint projects	$0	$310,300,000	$73,888,929	$1,401,100,000

*All values adjusted for inflation (2016 dollars).

state-specific information were employed either as a primary data source or to corroborate interview data.

Table 8.3 provides a comparative glance at the size and scope of the programs in the four case studies, from inception through FY 2019. New York's program is the largest in terms of cumulative federal assistance, total cumulative assistance, number of loans made, cumulative value of leveraged bonds issued, and CWSRF funds directed to nonpoint-source projects. Georgia has never operated a leveraged CWSRF program, but has contributed more from state-run (non-CWSRF) water programs than the other states. While three of the four states have funded projects under most of the categories of projects eligible for CWSRF assistance, Alabama has funded only centralized wastewater treatment works (WTW) for the life of the program. Georgia was one of the first states to receive an initial capitalization grant, while New York was one of the last states to receive its first capitalization grant.

Table 8.3 *State program comparison, case study states*

	AL	CA*	GA	NY
Year of first award	1989	1989	1988	1990
1990 population	4,049,000	29,950,000	6,507,000	18,000,000
1992 total needs	$848,000,000	$8,396,000,000	$1,944,000,000	$23,136,000,000
Needs per capita	$209.43	$280.33	$298.76	$1,285.33
Cumulative federal grant	$508,568,197	$3,375,634,918	$861,372,730	$5,367,234,295
Total state contribution	$112,820,041	$629,749,484	$156,753,405	$951,873,418
Additional state contribution	–	–	–	–
Total cumulative assistance**	$1,345,626,390	$10,996,968,647	$2,206,169,895	$17,208,653,677
Total # loans	274	838	422	1,727
Wastewater $**	$1,340,728,390	$9,531,580,821	$2,018,560,006	$15,968,894,904
Stormwater $**	–	$148,010,619	$34,735,666	$275,710,161
Energy conservation $**	–	$60,291,704	$2,708,000	$431,173
Water conservation $**	–	$984,074,221	$19,099,900	–
Nonpoint $**	–	$251,823,102	$131,066,323	$963,617,439
Other $**	–	$319,780,238	–	–
Gross leveraged bonds issued cum.**	$667,230,000	$1,609,960,000	–	$10,675,416,049
State-funded programs (sep from CWSRF) $**	–	$238,488,676	$518,717,570	$281,131,422
Federal ROI cumulative %	244	287	231	334

*California leveraged in only two years; 2003 and 2016.
**Dollar values given are not adjusted for inflation.

Case Studies

Alabama

In the early 1990s, Alabama had a population of roughly four million people, and water quality needs below the national average.[3] Alabama has operated a CWSRF program since the inception of the national program, making its first loans in FY 1989. Since the first year of operation, Alabama has made 274 loans totaling $1.346 billion, and has received federal capitalization grants totaling $508.6 million. The CWSRF program represents Alabama's only source of funding for water quality infrastructure; Alabama has not, in at least the last forty-five years, operated any state-funded programs to finance water quality infrastructure. To the extent water quality needs have been met in the state, they have been met through federal programs[4] such as the construction grants program, through community self-financing, or through private sector funding and ownership.[5] Thus the state has taken a very low-key role in water quality finance, providing no state funds for water quality above and beyond the required matching funds for the CWSRF capitalization grants.[6] To date, 100 percent of Alabama's CWSRF assistance has been directed to centralized wastewater needs; the state has not funded any stormwater, conservation, nonpoint, or estuary projects due to statutory limitations.[7]

Organization and Administration of the CWSRF Program

The development of Alabama's CWSRF program followed a pattern very different than that of many other states. When the US Congress passed the Water Quality Act in 1987, Alabama was caught in a period of uncertainty, since nearly all of Alabama's water quality infrastructure resources came from the federal government via the construction grants program. A private sector investment banker, sensing a chance to shape and develop a business opportunity, used his contacts in the state legislature to shape and enact the enabling legislation for the Alabama CWSRF program. Working with select members of both houses of the legislature, the investment banker drafted the legislation, lobbied members of the legislature, and became the force behind the bill. As one state official said, "[he] selected Alabama; Alabama didn't select [him]."[8]

The legislation was seemingly simple and innocuous. It created a "shell" organization, the Alabama Water Pollution Control Authority (AWPCA), but did not authorize any funds for the new authority. However, the legislation did give the AWPCA the authority to issue state general revenue bonds, make loans to communities, and assume administrative control of the CWSRF program. The AWPCA consisted of five formal members, all of whom were public officials – the Governor, the Lieutenant Governor, the Speaker of the House, the State Finance Director, and the director of the Alabama Department of Environmental Management (ADEM). The authority was constructed in such a way that the governor would always control a majority of members, since the finance director and the director of ADEM were gubernatorial appointees.[9] The enabling legislation gave AWPCA all formal program authority, including the authority to make decisions about the allocation of loans. The Facilities Construction Section of ADEM, formerly the WTW construction grants program office, provided administrative support in terms of maintaining a priority list of applicants, review of applications and plans, and general program administration.

The legislation also specifically limited use of CWSRF funds for centralized wastewater treatment projects. This feature was important in at least two respects. First, this provision in the law denied the use of CWSRF funds in the state to address nonpoint pollution, even though the state contains more than 132,000 miles of rivers and streams, and drains water from four states.[10] Because there were no state-funded water quality programs, the net effect of this was that

Alabama had no state programs to address nonpoint pollution. Second, by limiting CWSRF funds to centralized wastewater treatment, this virtually guaranteed a need not only for leveraging but provided a steady stream of revenue to the private sector actors that effectively managed the program.

The authority to issue state revenue bonds was a key requirement for the financial strategy envisioned by the investment banker, a strategy that called for aggressive leveraging of the CWSRF. The creation of a shell organization was politically important, in that it defused initial resistance from local bond interests who might otherwise have felt threatened by the creation of a statewide bond system. The strategy was apparently successful, in that interview respondents reported very little initial resistance from the local bond investment industry.

Because the authority had no funding or staff it relied on formal contracts with four different private sector actors – a trustee bank, a financial advisor, a bond attorney, and a bond underwriter – to operate the program. The formal role of the trustee bank was to handle the administrative tasks related to program funds: track repayment streams, disburse loan installments and interest payments on bond issues, and monitor investment growth on reserve accounts. The bond attorney advised the authority on the legal aspects of the bond issues, and prepared all paperwork and contracts related to the issuance of bonds. The bond underwriter assisted the bond attorney and the trustee bank in the execution of their duties, and addressed the legal and technical aspects of the bond insurance requirements. The financial advisor, working under contract, provided investment advice to the authority on matters relating to the financial content of the program.

There was a significant difference between the formal authority specified in the enabling legislation and the actual exercise of authority in the program. Although the AWPCA was given nearly all formal program authority under the legislation, two actors – the financial advisor and a state official in ADEM – made nearly all program decisions. Of these two, the more important actor was the private sector financial advisor. As previously noted, the financial advisor was the person responsible for the passage of the enabling legislation and the design and development of the program. In many respects, the financial advisor *was* the CWSRF program in Alabama. Although given nearly all formal program authority, the real role of the AWPCA was reduced to what one state official described as "benign neglect."

The financial advisor worked in tandem with a state official in ADEM to administer the program. ADEM handled the environmental aspects of the program – setting the priority list, reviewing plans and specifications, and conducting environmental impact assessments. Although environmental in nature, the functions controlled by ADEM in the CWSRF program were routine and technical, and required little in the way of decision-making authority in terms of the allocation of program resources. The financial advisor thus assumed many functions, including much of the real decision-making authority in the CWSRF program. The financial advisor created the bond pools, evaluated the creditworthiness of the applicants, negotiated the bond pool terms with the bond raters, set the loan terms and amounts, and tracked the financial aspects of the program.

The financial advisor also assumed the authority to structure the financial setting in the CWSRF program. In order to provide loan assistance at rates lower than commercial loans, it was necessary to assure that the bond rating on the pool of applicants would be as high as possible. The environmental priority system provided the primary means to organize the pool. However, in order to raise the bond rating of the pool, the financial advisor rated each applicant on the priority list in terms of financial soundness – the ability to repay a loan. If the potential pool contained any financially at-risk communities, the financial advisor attempted to recruit "elephants" – financially sound, large communities with high bond ratings – for inclusion in the pool. Although the "elephants" were required to appear on the priority list, their position on the list became irrelevant, as they were offered assistance on the basis of their financial position, rather than their relative environmental need. The Alabama CWSRF thus informally targeted financially sound communities as necessary to raise the bond rating of the leveraged bond pool. The recruiting (or, as one state official put it, "selling") of the CWSRF bond pool to potential "elephants" was left to the financial advisor, albeit with the tacit approval of ADEM.

The organizational culture and philosophy of the CWSRF program clearly reflected the importance of competition and the primacy of market forces and values. However, an area in which competition was conspicuously absent in the CWSRF program was in the decision to choose private sector consultants. As discussed previously, the primary private sector actor created the CWSRF structure in Alabama by literally showing up and taking the initiative to create a business opportunity in a vacuum of public policy non-activity. When asked, state officials spoke very highly of the talents and skills of the financial consultant, yet they deflected questions, and become defensive, about the lack of competition for the role of financial consultant. One respondent went as far as to say there was no need for competition, since the current consultant was responsible for the success of the Alabama program.

Financial Structure

Alabama engaged in an aggressively leveraged financial structure for the CWSRF program. In fact, Alabama floated bond issues every year to meet the required 20 percent state match for the federal capitalization grant. The state legislature has never appropriated funds for the state match, leaving CWSRF officials no option but to issue state revenue bonds to raise the state match. The financial structure used by Alabama was complex, and involved several revenue streams, reserve funds, and bond repayment funds. In essence, the federal capitalization grant was used to provide administrative costs and to capitalize a bond repayment account. Any additional money from the capitalization grant was placed in the master CWSRF fund, and was used to provide direct loans to certain classes of applicants.[11] As of FY 2019, roughly $1.345 billion had been disbursed in the form of direct loans.

In spite of the apparent efforts to maintain the financial health of the CWSRF program, state officials expressed some doubt about the long-term viability of the fund. When asked about the effects of offering loans below the prevailing rate of inflation,[12] a state official responded that inflation is not accounted for in the Alabama CWSRF model. When pressed for a reason why the effects of inflation were not addressed, the official replied, "we don't know how to predict the future – we'll worry about that later." The same general attitude about the future financial health of the fund was evident in a broader discussion of the long-term viability of an aggressively leveraged CWSRF financial structure. As one state official said, "we've sold out the future" for present needs. Although the state official thus agreed that aggressive leveraging has long-term negative consequences for the CWSRF program, he cited ambiguity in both the Water Quality Act and in Environmental Protection Agency (EPA) documents about the meaning of "perpetuity." "Nowhere in the CWSRF program," – he said, "in the EPA regulations, initial guidance, the [national] legislation, or other documents – does it give a definition of 'perpetual.' We're not sure what this means."

In sum, Alabama was a state for which the programmatic assumptions for initial state capacity were questionable. Although Alabama had operated the state component of the construction grants program, Alabama assumed primacy in the water quality program later than most states.[13] ADEM was conceived and structured as an implementer of federal environmental policy; the lack of state programs and a general lack of political support for environmental programs meant that the state lacked the administrative capacity to assume responsibility for a complex environmental program. Faced with this lack of capacity, Alabama was largely unprepared to implement the CWSRF. The preeminent role played by the private consultant in the development of the legislation, the structure of the state's program, and the use of leveraging to obviate the requirement for a legislative appropriation to meet the state match requirement are all indicative of this point. These decisions are not without consequences; as Morris (1994, 1997, 1999a) reports, a higher level of involvement of private sector actors in the CWSRF program fundamentally alters the distribution of program resources. The recruitment of "elephants" – large communities with strong bond ratings – in bond pools are evidence of this effect. While Alabama continues to leverage and to issue loans, the fundamental administrative structure of the program is largely as it was at the point of initial implementation.

California

In 1990 California was home to just under thirty million people, making it one of the most populous states in the nation. With such a large population California also had significant water quality needs, reporting total needs of approximately $8.4 billion in the 1992 Needs Survey. The development of the CWSRF program in California reflected several relevant factors that resulted in a decision not to leverage their capitalization grants. In addition to a high level of needs, the state also possessed significant experience with state-run programs designed to meet water quality needs. Officials in the state examined the leveraging question carefully and ultimately decided that leveraging was not worthwhile. Since the US Congress and EPA were actively encouraging states to leverage, California presents a case of a state with large needs that initially chose not to leverage.[14]

CWSRF History and State Context

California has traditionally been in the forefront of state efforts in environmental protection. For example, the state led the nation in legislation to protect air quality, and California has typically created state environmental protection requirements that have exceeded those required under national law. In 1967 the California legislature passed the Porter-Cologne Water Quality Control Act that, among other things, created the State Water Resources Control Board (SWRCB) and created state-funded water quality programs. California moved quickly to develop its CWSRF program; the governor signed Section 1, Chapter 6.5 of the Water Code of the State of California in late September 1987, thus creating the California State Water Pollution Control Revolving Fund. Appropriations under the program were allocated to the SWRCB for expenditure. The SWRCB consisted of five members appointed by the governor and was empowered to protect the state's water quality and to allocate water rights. The SWRCB worked with EPA to administer the construction grants program starting in 1972, and ultimately was given authority to administer all of California's programs related to the Clean Water Act, as well as four state-run and state-funded programs. The first of these programs was the Small Communities Grant Program, initially funded by a $25 million state bond issue. The program provided planning and construction grants of up to $2 million, and was targeted to communities with a population of less than 3,500 residents.

The other three state programs were loan programs. The Water Reclamation Land Program was designed as a revolving loan fund program. Under the law, 50 percent of the loan repayment stream is returned to the fund, while the other half is directed to the state's General Fund. Experience in the operation of this fund provided state officials with experience in the operation of revolving loan funds, which proved invaluable with the advent of the CWSRF program. The Agricultural Drainage Water Management Loan Program was capitalized with $75 million, and provided twenty-year loans at a subsidized interest rate. The final program was the Water Quality Control Loan Program, a $6 million fund intended as a "last resort fund" to provide loans of up to twenty-five years to communities that demonstrated financial hardship. The program also required a local election for voter approval of the amount of the loan. California thus had in place several state-run and state-funded water quality programs, including a revolving fund, and had established substantial state capacity to design, administer, and manage complex water quality programs. However, these state programs were clearly not sufficient to meet the high level of water quality needs in the state.

California officials had also participated in early discussions regarding a national CWSRF program. As detailed in previous chapters the EPA began discussions in the mid-1980s to determine how the federal government could move beyond a categorical grant program and shift responsibility for funding and program administration to the states. Based on these discussions, state officials were confident that federal water quality programs would shift from a grant-based program to a loan program. In preparation for the change, the state began aggressively to address its water quality needs, in order to be in a position to assume full responsibility for the new CWSRF program. By FY 1989 California had cleared its backlog of projects on the EPA's National Municipal Policy (NMP) list. The state was able to certify that all projects on the NMP list had either been funded or were already on enforceable schedules.

State program officials also determined early on that substantial problems would likely accompany the transition from a grant to a loan program if the state were to operate both programs simultaneously. A decision was thus made to apply for the initial capitalization grant as quickly as possible, and to transfer from the grants program to the loan program as rapidly as practicable. The target date for completion of the transition was June 1988; although the state still had a large number of grants projects still in progress, the final grant authorization was made in 1990. The staff of the construction grants program became the CWSRF staff. When the last project under the construction grant program ended, so did the state's delegation agreement with EPA. The state then assumed full responsibility for the CWSRF program, as well as the preexisting state funding programs. California officials thus saw the CWSRF program as a mechanism through which the state could improve program administration, reduce federal restrictions, and strengthen the role of local governments, while establishing a perpetual fund for a state administered water quality program.

CWSRF Organization and Administration

The 1987 revisions to the California Water Code established the state CWSRF program and empowered the SWRCB to make loans, guarantee or purchase insurance, provide revenue security for payment of municipal bonds, refinance municipal debts, and guarantee loans for revolving funds established by municipalities, for water quality projects that met eligibility criteria. With the switch from the grants program, twenty-six staff members from the construction grants program were assigned to the CWSRF program. The state's CWSRF director was chief of the Management Support Section, and a senior SWRCB engineer. The CWSRF staff included personnel from both the construction grants program and other state-administered water quality programs. Other state offices also have formal CWSRF program responsibilities, including the state treasurer, the state comptroller, and the Department of General Services. These offices assumed centralized functions such as approving contracts, disbursing payments, and selling bonds. The California CWSRF organization was primarily a single agency structure although several functions are allocated to other state agencies. The CWSRF was administered similarly to other state water quality programs. The administrative structure was developed by a subcommittee of the Trilateral Technical Advisory Committee (TRITAC), a technical advisory committee component of the SWRCB. TRITAC predated the development of the CWSRF program and was composed of representatives from the California Water Pollution Control Association, the League of California Cities, and the California Association of Sanitation Agencies.

The California CWSRF structure was developed and shaped by state environmental personnel, based on existing programs and previous experiences. California officials reported a close working relationship with both the EPA Region 9 office and headquarters offices, but the state restricted the influence and role of both entities as much as possible during the development of the CWSRF. The high level of experience and knowledge of SWRCB staff, coupled with a number of self-studies and a desire to be independent, meant that the state rarely consulted with EPA. EPA did provide initial guidance (as it did to all states) and responded willingly to inquiries about the CWSRF program. From the state's perspective, the role of EPA was to approve finished products sent to them by the state.

Financial Structure: CWSRF without Leveraging

The California CWSRF is structured, by and large, as a straight loan program. During the transition from the construction grants program to the CWSRF program, the SWRCB initially recommended a legislative approach that would allow for leveraging to increase the capital available for CWSRF loans. The proposal was largely opposed by local governments, the state treasurer, TRITAC, and others. Many medium-sized and financially sound municipalities suggested that large communities were capable of self-finance options, and they objected to subsidizing financially weaker communities. The state treasurer opposed the legislation because it would allow the SWRCB to issue bonds. The state employed a private sector financial consultant who was retained in an effort to solve these problems, but the effort was unsuccessful. According to the CWSRF coordinator, the root of the problem was a combination of political and structural factors.

The SWRCB ultimately concluded that leveraging was not only unnecessary in California, but would actually be harmful to the CWSRF program. The reasoning was as follows. Although leveraging could provide an increased amount of initial funding, there was concern that funding would be reduced in later years.[15] In an interview the state's CWSRF coordinator pointed out that leveraging serves to exacerbate the "up and down cycle" of program funds; the net result would generate significant problems for stable staffing of the CWSRF program. The CWSRF coordinator was also concerned about the instability leveraging would create for municipalities, and the ability of the state to maintain a good working relationship with these communities. As structured, the CWSRF does not pay for planning costs related to a funding request unless the proposal is funded. Without the assurance of a stable program, the fear was that many communities would simply choose not to participate. The program depended on the inclination of local officials to "buy in" to loans; although the level of need in California is high, the level of demand is not as strong. The success of the state-funded program affected demand for major projects that might otherwise be appropriate for CWSRF funding. Although a number of large coastal cities had a need for funding, the projects were all much larger than the CWSRF program could reasonably fund without leveraging. The CWSRF coordinator indicated that if those large cities decided to proceed with their projects, the state would likely reconsider the leveraging decision. As long as the state could meet existing demand with a straight loan program, the state would shelve plans to leverage. Eventually California leveraged in two years during the period of study for this book; in 2003 the state raised $314.1 million in a bond issue, and in 2016 the state issued another $500 million in bonds to support the CWSRF program.[16]

In sum, California's extensive experience with state-run water quality programs, strong political support, a clear picture of the state's needs and the needs of its communities, a concern for the long-term effects of leveraging, and a strong desire for the state to be the preeminent actor in water quality all shaped the development and operation of the state's program in the early years. California continued to operate a simple loan program until 2003, when the state adopted leveraging for the first time. The decision was influenced by several factors, including growing water quality needs fueled by a growing population, aging infrastructure, and state budget shortfalls. Still, the California program is administered fully by state program personnel. In this case, the presumption of adequate state administrative, financial, and political capacity was clearly met.

Georgia

The 1990 census reported a population for Georgia of about six million people. Although Georgia is primarily a rural agricultural state, its capital, Atlanta, is a regional transportation and business nexus, and has experienced a great deal of growth in recent years. Georgia's water quality needs were slightly below the national average, totaling about $1.94 billion in the 1992 Needs Survey. In addition to, and preceding, the CWSRF program, Georgia operated a sizable state-funded program to finance water quality infrastructure needs. Administered by the Georgia Environmental Financing Authority (GEFA), the program provided loans to communities for a variety of environmental infrastructure needs. The amount of GEFA assistance was typically less than the assistance provided through the CWSRF program,[17] but the GEFA program represented a commitment on the part of the state to address water quality infrastructure needs with state programs as well as federal programs.

Georgia was among the first states to adopt the CWSRF program. Indeed, the enabling legislation to create the CWSRF program in Georgia was passed by the state legislature in March 1986, nearly a year prior to the passage of the Water Quality Act in 1987. State officials in Georgia cited a very good working relationship with the EPA's Region IV office in Atlanta as a major factor in the early passage of the legislation. Region IV personnel acted as intermediaries between EPA headquarters and state officials in Georgia. The result was the passage of a short (three paragraphs), somewhat vague bill that gave authority to state officials to create and administer a revolving loan fund program, pending the outcome and details of the national legislation.

One of the key points cited by state officials was that passage of the legislation in Georgia was made smoother by the fact that the legislature did not need to appropriate additional funds to the CWSRF program. A provision of the Water Quality Act allows states to use "in-kind" matches for the federal capitalization grant, and Georgia received the approval of EPA to apply state loans made through GEFA to the match requirement. A more practical reason was a perception on the part of state officials that if the legislature needed to appropriate money directly to the CWSRF program, their response would be to take money away from GEFA and give it to the CWSRF program. State officials in both the Department of Natural Resources (DNR) and GEFA agreed that such a move would not be in the best interests of either the state or the environmental goals of the state. State officials, with the assistance of the Region IV office, negotiated the use of an in-kind match that allowed Georgia to receive the federal capitalization grant without the expenditure of additional state funds.

Georgia's first capitalization grant was awarded in March 1988, and the first loans executed some months later. Georgia had a number of applicants ready to proceed with water quality infrastructure construction, so there was no shortage of applicants for CWSRF assistance. In fact, Georgia had more applicants for assistance than funds available for the first several years of the program, although one state official indicated that demand waned in the early 1990s as market interest rates began to fall. CWSRF officials in Georgia reported that communities were generally receptive to the idea of a loan program. CWSRF program staff did not undertake any aggressive marketing programs, although one state official regularly attended local government professional and association meetings (such as the Georgia Municipal Association meeting) to provide information about the CWSRF program. The marketing efforts were very informal. As one CWSRF official reported, "we didn't do any dog and pony shows."

The goal of attracting potential applicants to the CWSRF program in Georgia was addressed through certain program incentives to reduce the costs, both monetary and administrative, to communities. State officials proudly referred to a reputation of "being easy to work with." As a part of the underlying program philosophy, state officials placed great emphasis on mechanisms designed to reduce waiting time, minimize the burden of federal cross-cutting requirements, simplify the application for assistance, and provide technical assistance and financial consulting to any applicant requesting help. One state official noted the decreased administrative burden of the CWSRF program and the speed at which construction can begin, as compared to the construction grants program: "The extra speed is important . . . we're building something, we're not sitting around shuffling papers. We're here to build treatment plants, not to cause problems." Although the availability of below-market loan rates also serves to attract applicants to the CWSRF program, CWSRF officials still believe a "user-friendly" administrative structure is a necessary component of a good program. As one state official said, "If you make it extremely difficult they won't come to you, no matter how cheap the money is."

CWSRF Administrative Structure

Up until July 1, 1994, the CWSRF program in Georgia was administered jointly by the Environmental Protection Division (EPD) of the Georgia Department of Natural Resources and the Georgia Environmental Financing Authority. Under that system, EPD assumed responsibility for many of the programmatic decisions made in the CWSRF program. EPD processed applications for assistance, reviewed design plans and specifications, maintained the priority list, performed environmental assessments, and tracked loan repayments. GEFA officials assumed mostly coordinative activities, including a financial review of applicants, monitoring loan accounts, and preparing loan materials for execution. Loan agreements were signed for the state by the director of the EPD and the commissioner of GEFA.

Most program and decision authority rested with the CWSRF coordinator in EPD. As administrator of the program, the CWSRF coordinator bore responsibility for a small staff, consisting mostly of environmental engineers, and for program coordination with officials in GEFA. The CWSRF coordinator exercised responsibility for the implementation of the program by coordinating the activities of CWSRF personnel in both EPD and GEFA. Based on the recommendations of the staff and the successful completion of the technical requirements of the application (design review, financial review, environmental impact assessment, etc.), the CWSRF coordinator forwarded a letter to the signatory authorities in both EPD and GEFA recommending execution of the loan agreement. After review of the supporting materials, the signatories would execute the loan agreement, and construction would proceed.

A new administrative mechanism was implemented in July 1994. The most obvious change in the CWSRF structure was that programmatic authority for the CWSRF program shifted from EPD to GEFA. Accompanying the shift in authority was a change in the duties of both EPD and GEFA. EPD retained responsibility for the environmental/technical aspects of the program (design review, environmental impact assessments, etc.), and continued to track loan repayments for the program. EPD also continued to monitor the construction grants program[18] and monitored loans made up through FY 1993.

The changes for GEFA were more substantial. As a means of consolidating program authority, GEFA assumed program authority for both the state-funded loan program and the CWSRF program. According to a state official, the primary force behind the reorganization was a desire to centralize funding authority. In order to better coordinate the funding decisions of both GEFA and the CWSRF program, a decision was made to transfer the program authority to GEFA. In many ways, this shift was representative of the inherent tensions between the environmental goals and the financial goals in the CWSRF program. As one EPD official indicated, the financial requirements of the CWSRF program fundamentally altered the administrative approach required to administer a successful program. "We're not a bank," the official said, "but we often have to act like one." By transferring program authority to GEFA, the CWSRF program gained an experienced and successful state agency to administer the financial portions of the program, and to allow in-house coordination of the CWSRF program with the state-funded program administered by

GEFA. State officials expressed a prediction that the shift in program authority to GEFA would allow a higher degree of coordination between the two funding sources, thus more effectively serving the interests of the people of Georgia.

The shift in program authority carried with it minor changes in the financial structure of the CWSRF program. The state continued to use GEFA loans as an in-kind match, but the interest rate charged on CWSRF loans was more closely tied to the rates for state general obligation (GO) bonds. The rates for CWSRF loans were projected to be slightly lower than the GO bond rate, a move designed to counter the additional costs to communities of the cross-cutting federal requirements that were an integral part of the CWSRF program. In addition, GEFA developed plans to undertake a more active marketing effort for the CWSRF program, combining CWSRF marketing with a broader-based plan to market financial assistance to communities in Georgia.

An additional feature of the interagency agreement between GEFA and EPD was the availability of EPD's technical expertise and administrative authority to GEFA in its CWSRF activities. Rather than duplicate the administrative and technical authority of EPD, GEFA shared the necessary authority with EPD. EPD continued to maintain much of the design and technical authority in the program, but GEFA, through its assumption of program authority, could draw on EPD's technical and administrative authority to administer the CWSRF program more efficiently.

Financial Structure

Georgia operated a direct loan program. Loans were generally made at a fixed rate of 4 percent, although the loans were made at a 2 percent rate for the first few years of operation. Although early demand for CWSRF assistance was substantial, Georgia has never operated a leveraged program. In the early years of the program, the investment banking industry, both local and national, tried very hard to convince state CWSRF officials to leverage the program. State officials resisted the pressure for two reasons. First, state officials felt that while early demand was high (three or four times as much as the available program resources), the combination of CWSRF loans and GEFA loans would cover the demand. Second, state officials were very concerned about the effects of leveraging on the long-term solvency and productivity of the fund. State officials were convinced that while leveraging would provide a greater pool of money in the near-term, the long-term effects of leveraging would significantly degrade the value of the fund over time. As one state official said at the time, "we felt that aggressive leveraging is a sure way to *not* have a fund in perpetuity. We came to the conclusion that it was not in the best interests of the state to leverage."

In summary, Georgia employed a direct loan strategy for the CWSRF program. In addition, Georgia used loans made through a state-run and state-funded infrastructure financing program to meet the requirements for CWSRF matching funds. The loan applicant paid a 4 percent loan origination fee to help defray future administrative costs. The net result has been a program that was able to employ existing state agencies, both administrative and financial, for program administration, and one that enjoyed a significant level of political support. The fact that Georgia had its state enabling legislation in place eleven months prior to the eventual passage of the WQA illustrated this level of support. Moreover, the decision to allocate GEFA funding to meet the 20 percent match requirement also meant that Georgia avoided a potentially difficult political battle over funding.

The other interesting element in the Georgia case is the early restructuring of the administrative arrangement for the CWSRF program. While program officials reported at the time that the initial structure was clearly adequate for the program, the additional administrative efficiencies gained through the reorganization, coupled with the simplified structure for applicant communities, reinforced the value of the restructuring effort. When the Drinking Water State Revolving Fund program was enacted in 1996, the same structure was employed for that program as well. Georgia possessed a high level of existing administrative capacity, and the state was able to make good use of that capacity both in terms of initial implementation and the later restructuring process.

Finally, Georgia has used CWSRF money to fund projects in all but one category of allowable uses: estuaries.[19] As is the case in most states, the bulk of CWSRF money has been allocated to meeting wastewater needs. However, Georgia has allocated nearly 5 percent of all CWSRF funds, or roughly $131 million, to nonpoint projects. By contrast, this represents over 50 percent of the funds spent by California, a state much larger in terms of both population and CWSRF program size, on nonpoint projects.

New York

New York is notable for a high level of water quality needs. The 1992 Needs Survey reported total state needs of roughly $23.1 billion, by far the most in the nation. By comparison, California, a much larger state in terms of both land area and population, had needs slightly over a third of those of New York. New York's population in 1990 was eighteen million; about 40 percent of which (7.32 million) were in New York City. New York was one of the last states to receive an initial capitalization grant. Like California and Georgia, New York also operates a state-funded and -operated water quality program administered through the New York Environmental Finance Corporation (NYEFC). Compared to the size of the state's water quality needs, the program was relatively modest in size; to date it has provided about $281 million in grants and loans, mostly to small communities. This meant that when implementation of the CWSRF began, New York already possessed a high level of financial expertise in state government. Moreover, as we will see, there was already a high level of coordination between the environmental and financial agencies in the state.

The experiences of New York and their CWSRF program are clustered around three central points. The first involves the specific tailoring of policy choices and instruments to meet certain kinds of need. As was the case in several other states, New York approached the development of the CWSRF program from a proactive point of view. This is particularly true in the case of decisions about leveraging. New York had a clear perspective regarding the role of the state in terms of aid to small communities, a theme that helped drive many of the early programmatic decisions. A second point is related to the issue of how New York addressed the need for personnel from two different agencies (environmental and financial) to work together to design, implement, and administer the program. The CWSRF program intertwines both environmental and financial elements, and New York made a concerted effort to inculcate the objectives and values of both perspectives in its organizational framework. This task requires careful management to succeed. The case of New York exemplifies a careful strategy of organizational consolidation. Finally, the third point is related to the marketing efforts required to turn the CWSRF model into an effective policy instrument. Rather than a "one size fits all" approach to marketing, New York officials realized that a successful marketing strategy must not only market to communities, it must also successfully market the bond offerings used to capitalize the program through the use of leveraging. These three points provide the story of a more fundamental lesson that has emerged from these case studies: The financial requirements of CWSRF programs tend to overshadow the environmental goals that the Water Quality Act tries to address. The tension between these two elements creates a substantial challenge (and opportunity) to the administrators in states to create a program that meets both sets of goals.

When it became clear in the mid-1980s that the construction grants program would be replaced with a loan program, the state governor provided support for the development of state legislation. However, an affirmative action issue delayed passage of the legislation until July 1989. Once the legislation was adopted the state agencies moved quickly to implement the CWSRF; draft rules and regulations were completed within two months, and in December 1989 the state filed its first capitalization grant application with the EPA, along with a plan for aggressive leveraging. The first federal grant was awarded in early 1990, and the state executed its first CWSRF loan in May 1990.

CWSRF Design and Implementation

Two of the primary agencies in New York entrusted with CWSRF design and implementation are the state's Department of Environmental Conservation (DEC) and the New York Environmental Finance Corporation (NYEFC). A critical element of the New York State case concerns how the state responded to the task of interconnecting financial and environmental management capacity. The state did engage external advisors to assist in the process, especially in the development of the state's leveraging model. But, the issue of public sector organizational arrangements (both capabilities and personnel) was prominent on the agenda from the early days of program design. The initial proposal for enabling legislation recommended that DEC serve as the lead agency for the program, but DEC had no statutory authority to issue bonds. Therefore, program designers decided early in the process of CWSRF design that, due to New York's massive water quality needs, leveraging was necessary to provide the maximum effect of the available resources. A public authority was required in order to issue revenue bonds, and the decision was made to enlist an existing entity with the statutory authority to do so – the NYEFC. Moreover, there was an existing organizational linkage between the DEC and EFC, in that the chairman of the NYEFC board was, by law, the commissioner of the DEC. In this arrangement, the DEC conducted technical and environmental reviews of applicant projects, and set the state's priority list for funding. The application was then sent to the NYEFC, who served as the recipient of the federal capitalization grant. The NYEFC then passed the application back to the DEC, whose personnel worked with the state comptroller to disburse and track loans.

In the early stages of the program staff from the two primary agencies established what an interviewee termed "a working relationship that worked." However, under the dual-agency arrangement, a variety of issues of varying complexity continued to arise, which in turn led to tensions between the agencies. As a result, both DEC and NYEFC began considering an administrative consolidation process. This culminated with the reassignment in April 1991 of forty-seven positions from DEC to NYEFC, a decision made by the DEC commissioner. There were a number of administrative issues that needed to be addressed, including which employees would leave DEC, whether legislative approval was needed (it wasn't), and how union representation would be achieved. The shift of personnel from DEC to NYEFC continued as the few CWSRF positions remaining in DEC became vacant.

Although the decision to consolidate CWSRF operations into the NYEFC was made at the top levels of the DEC, efforts to prepare for a smooth transition were carried out at lower levels of the organization. NYEFC scheduled meetings at which personnel from each organization could interact with each other and discuss the issues. An outside facilitator was brought in to engage the participants and ensure a worthwhile experience. The effect was that people from both groups learned a great deal about the "other side of the house." The facilitator made careful efforts to inspire the participants to focus on the program as a single program rather than as a function of two separate organizations. An executive committee was formed of participants from both agencies to address policy issues. A committee of senior staff was also created, again consisting of people from both agencies. The committee's task was to identify day-to-day coordination issues, and to find solutions to those issues.

Officials in New York State reported multiple advantages to the use of loan programs as policy instruments. One advantage was that loans provided an incentive to communities to reduce costs. Loan programs could also be tailored to the specific financial conditions of each applicant community. For example, interest rates were structured to provide a negative rate to the neediest communities. Likewise, different products could be attached to the needs of different applicants. The New York CWSRF operated a direct loan program in addition to an aggressively leveraged loan program. The direct loan program was designed primarily to meet the needs of financially at-risk communities, while the leveraged program was designed to generate a large fund corpus to meet substantial, immediate needs.

The effort to assist small and financially at-risk communities illustrated the matching of specific policy instruments to specific needs, and the complexities that arose as a result. The direct loan program was slow to develop. As of mid-1991only four direct loans had been issued. The principal reason for this slow rate of implementation was due to the demands placed on state resources by the much larger and more complex leveraged program. The second reason was entirely political; interviewees indicated mixed perspectives among the NYEFC Board members regarding the direct loans program. Some members believed that the environmental impact of the direct loans would be too small to justify the investment made; other members took precisely the opposite position.

The direct-loan program in the New York CWSRF represented one method to responding to the needs of very small communities within a program dominated by an aggressively leveraged policy instrument. Another element involved a program to promote self-help by communities. Beginning in 1986 the NYEFC had participated in a cooperative relationship with two state agencies and a nonprofit organization to provide technical assistance to very small communities in the state. The objective was to help them identify ways to reduce costs to make projects more affordable; in some cases, communities in the program were able to reduce project costs by as much as 50 percent. This technical assistance was time-consuming and difficult, but was also effective. One important source of support for this effort was EPA, who allowed the state to assume the costs of the program as long as the community was legitimately regarded as a potential customer.

New York's leveraging strategy reflected the two classes of need for water quality funding in the state. The first class was represented by the enormous funding needs of the New York City Water Authority; the other class was represented by the needs of other communities and local authorities in the state. The state employed a reserve fund approach in both classes, and also employed several levels of security for the bonds issued by the NYEFC. For New York City, the approach was to create a reserve fund equal to one-third of the amount required. The money in this reserve fund was used to support the bonds issued to cover the loan. As the loan principal and interest were repaid, a third of the repayment stream was directed to the reserve fund. Over time, these funds were available to be revolved to new loans.

For other communities in the state, the NYEFC formed bond pools to provide financing assistance. The bond pools were secured through several mechanisms. First, state law allowed the NYEFC to ask the New York State controller to deny aid to delinquent communities; that aid could then be transferred to the CWSRF to repay the loan. Second, applicant communities issued general obligation bonds rather than revenue bonds, which served to enhance the credit rating of the bond pools. Third, all borrowers except New York City were subject to a financing indenture. As communities made principal payments and funds were released from the reserve fund, a review of all the pooled communities was made. If the debt service reserve of an individual participant did not meet the one-third requirement, then the state brought it up to that level.

The ability of the state to market CWSRF bond issues was closely tied to the creditworthiness evaluations made by bond raters such as Moody's, Fitch, and Standard and Poor. The rating agencies could take varied approaches to their evaluation, an issue the state addressed. For example, the bond rater could base the rating on the creditworthiness of the lowest-ranked community in the pool, or base the rating on the overall strength of the pool and the expected cash flow from the pool. If the rating agency was willing to adopt the latter approach, then the ability of the CWSRF to serve financially weaker localities in the bond pools was improved. Because this approach could enhance program outcomes and allow the state to better serve small and financially at-risk communities, the state went to some lengths to encourage bond rating agencies to accept a broader view.

The case of New York represents a large state with significant water quality needs. The state realized early that the shift from grants to loans would require a new structure and a different administrative approach, but the state's experience with its own loan programs meant that the transition was straightforward. Moreover, the two state agencies involved made concerted efforts to make the transition as smooth and seamless as possible. The result was a program that offered a mix of direct and leveraged loans, since state officials concluded that such an approach would allow them to best serve the two very different classes of communities in the state. In this instance, the WQA's assumption of state capacity was clearly met, even if it did take about two years for the state to apply for its initial capitalization grant.

The New York case is also notable for its attention to marketing the CWSRF program, not only to applicant communities, but also to the bond market. By actively engaging the bond market, the state was able to lower the cost of assistance to communities, and ultimately better serve small and financially at-risk communities. Where Alabama officials had to recruit "elephants" with high bond ratings to form pools with reasonable bond ratings, New York's approach allowed much more latitude in the construction of the bond pools, and thus assured that environmental need was not entirely swamped by financial considerations. New York was thus able to strike a reasonable balance between the two.

Discussion

The four states discussed in this chapter represent very different implementation strategies of a national policy and program. These case studies speak to the question of whether or not a centralized (national) policy can be decentrally implemented and yet meet national policy objectives. Designed to allow a great deal of state discretion in the design, implementation, and administration of the program, the CWSRF policy instrument has fulfilled that design criterion. States have approached the issue of CWSRF design and administration in very different ways, and these four states are representative of these differences. The choices made by states in the early years of the program have determined, in large part, the program outcomes detected thirty years later.

Alabama represents a case in which the private sector has been included in the CWSRF institutional arrangement as agents of state government to administer the CWSRF program in the face of a clear lack of state capacity to design and implement the program. However, the fact that a private sector agent effectively designed an institutional arrangement to reflect private sector values, rather than public sector values, results in the CWSRF "good" – the loan – being transformed from a public good to a private good. In contrast, the interest of the other three states in the maintenance of public values, goals, and control resulted in the CWSRF loan being treated as a public worthy good, available for the benefit of all citizens of the state (within the constraints of the revolving loan fund model). Furthermore, the Alabama CWSRF institution was structured such that a private sector agent assumed an important role in the sovereign provision function of government – the ability to decide who receives CWSRF assistance. The authority to assume the provision function was also clearly a result of informal privatization. All formal authority in the

Alabama CWSRF was held by a political body (AWPCA). However, the AWPCA did not exercise its authority; rather, the private agent assumed the provision function.

Compared to the other cases presented, national policy goals and public accountability in the Alabama CWSRF program were threatened. While a public sector official in ADEM also functioned as an agent of AWPCA, the role of the public official was effectively limited to a repetitive, technical, support role for the private sector actor. It was not clear that the inclusion of the ADEM official provided an effective accountability mechanism. This situation was consistent with private sector values, in that accountability was determined by profitability and performance. In the case of Alabama, the CWSRF program was designed to maximize the near-term use of program resources, and to do so as efficiently as possible. Private goals were thus incorporated into the institutional structure of the Alabama CWSRF, which limited the ability of the program to address some public goals, especially when the public goals were in direct conflict with the private goals and values. In Alabama, the inherent tensions in the CWSRF program between financial and environmental objectives were evidenced by privatization of core elements of the program, including the sovereign provision function of government. By contrast, California purposefully kept its program structure straightforward, and implemented the program using public sector officials.

Georgia and New York represent cases in which program authority and accountability (provision) were retained by the public sector. These programs were designed and implemented by public officials interested in (1) achieving public goals, and (2) maintaining the status of clean water as a public good. Program authority was exercised by public officials, and the exercise of that authority matches the formal distribution of authority specified in the enabling legislation. While this approach did not necessarily negate the tension between the financial and environmental goals of the program, it did allow the state to assert clear public control over program resources, and thus enhance program outcomes.

The implications for the abilities of these states to meet the needs of small and financially at-risk communities were, in large part, due to the institutional arrangements employed, and the values represented in those arrangements. The inability of the Alabama CWSRF program to serve financially at-risk communities is largely a consequence of the use of an aggressive leveraging technique that required participating communities to be financially sound, coupled with a private sector-like approach to creditworthiness. On the other hand, Georgia and New York adopted financial strategies that not only met state requirements, but that allowed their respective CWSRF programs to serve the needs of at-risk communities, at least in a limited manner. California's early decision to forego leveraging not only fits the needs and demands at the time, it also allowed California to maintain flexibility in the program. As demand changed in later years, California was able to pivot to a leveraged program to meet the increased demand.

Comparing Outcomes in the States

A national survey of state CWSRF coordinators was conducted in 1990 by Heilman and Johnson (1991).[20] The survey consisted of 45 questions, and nearly 200 discrete data points. The survey covered a wide range of topics related to CWSRF design, structure, implementation, administration, and program outputs. Several of the responses from the four case study states are presented in Table 8.4. Of particular interest for this study are the factors that were important in the state that shaped the nature of the CWSRF program at initial implementation, and the sufficiency of state resources to assume responsibility for the program.

The two factors that were consistently reported as a "major influence" for the four case study states was the ability to meet environmental needs. This finding was not particularly surprising; the four respondents listed here were all long-term employees in a state environmental agency, and the WQA was very much an

Table 8.4 *Survey responses: Important factors and initial resource sufficiency, case study states*

	AL	CA	GA	NY
What factors were important in shaping your SRF?				
Meet environmental needs	Major	Major	Major	Major
Build state administrative capacity	Minor	None	Major	Major
Increase state flexibility/discretion	Moderate	None	Major	Major
Build long-term state financial capacity	Major	Major	Major	Major
Need to secure funding for water quality	Moderate	Moderate	Major	Major
Did governmental agencies in your state have sufficient resources to assume the primary role in the water quality area at the time of your first CAP grant?				
Political	Yes	Yes	Yes	Yes
Organizational	Yes	Yes	Yes	Yes
Managerial	Yes	Yes	Yes	Yes
Budgetary	No	Yes	No	Yes
Which of the following describes the overall approach your state took in implementing your SRF?	Neither	Rapid	Rapid	Slowly
Do you have a marketing program for your SRF?	Yes	No	No	Yes
Do you operate a state-funded water quality program?	No	Yes	Yes	Yes

environmentally focused law. However, the element on which there was agreement was the need to secure a long-term funding source to meet water quality needs. Although three of the four case study states operated state-funded water quality programs, none of these programs was large enough to meet the state's water quality needs. In interviews conducted in 1991 and 1994, the state coordinators expressed the sentiment that the national water quality goals could only be met with federal assistance. Still, the agreement about the importance of both the environmental and financial goals speaks to the inherent tensions within the CWSRF model. The interviewees all expressed concern about the long-term viability of the funds, particularly leveraged funds. Two of the interviewees felt that the planned federal capitalization funds would be inadequate to meet state needs,[21] and they worried about the effects of a lack of adequate funding on environmental quality.

The responses of the coordinators in the case study states in terms of the desire to build state administrative capacity were consistent with the interview data. Georgia and New York reported this as a major factor, and both states redesigned and reorganized their CWSRF administrative structures within a few years of initial implementation. In both cases, the state administrative structures emerged from the reorganization with enhanced administrative capacity. Alabama reported a desire to build administrative capacity as a minor factor, and the CWSRF structure reflected this response. The new "structure" consisted of a board of five elected and appointed officials that effectively played no substantive role in the program. The financial components were outsourced, and the ADEM staff did, in effect, the same job they had done under the Title II program. Finally, California reported this factor as no influence. California has built one of the strongest environmental programs in the nation; coupled with experience with state-funded water quality programs, California had no reason to see the CWSRF program as an opportunity to build state administrative capacity.

The question of the ability to increase state flexibility and discretion also showed variation across the cases. For California, a state somewhat notorious for its application of cross-cutting requirements, the federal cross-cutting requirements were not seen as onerous. In addition, California had always enjoyed a good working relationship with both the Region 9 EPA office (located in San Francisco) and EPA headquarters, and felt that their needs were generally met. The somewhat surprising case was Alabama. An important element of the political culture in the state was the importance of states' rights, and a federal program that effectively enhanced states' rights should have been a popular choice for the state. In an interview, the director of ADEM commented that "the elements of the program were already set, and we were more interested in developing a workable program."

In terms of sufficiency of resources, all four case study states generally reported that initial resources were sufficient. The exceptions were the two southern states, Alabama and Georgia, who each reported that budgetary resources were not initially sufficient. Program officials in Alabama knew there was no chance of securing a legislative appropriation for the state match requirement, which was one of the factors that led them to develop an aggressively leveraged program. In Georgia, initial uncertainty over the status of the state match, and the initial concern that the match funds would be reallocated from the state-funded water quality program, led Georgia officials to scramble to find a suitable solution. The Georgia CWSRF coordinator reported that the solution to apply existing funds from the state-funded programs as an in-kind match was made possible because of a close working relationship between the state and the Region 4 EPA office (located in Atlanta).

Differences in implementation speed were consistent with the interview data. As noted earlier in this chapter, Georgia passed its enabling legislation nearly a year before the final passage of the WQA, and they were among the first states to receive a capitalization grant. California followed a similar pattern; they viewed an unused capitalization grant as "money left on the table." Alabama reported they moved neither slowly nor rapidly, although this response likely reflects a respondent bias. ADEM was largely uninvolved in the development of the program structure and the enabling legislation; this was spearheaded by a private banker working directly with members of the legislature. In fact, Alabama enacted its enabling legislation four months after passage of the WQA and applied for its initial capitalization grant later that year. New York reported they moved slowly, but the "slowly" element was the legal tussle in the state legislature regarding an affirmative action issue that was tangential (at best) to the program. After the legislation was passed, New York moved very rapidly to implement its program. Still, the delay meant that New York was among the last states to apply for a capitalization grant.

Finally, the question regarding marketing programs for CWSRF assistance was also instructive. The non-leveraged programs did not report a marketing program, while the two leveraged states (Alabama and New York) developed and implemented marketing programs. The increased costs of a loan associated with leveraging required leveraged states actively to seek applicants for loans. It is important to note that when California adopted leveraging in 2003, they also implemented a marketing program.

Summary

These four case studies illustrate the importance of state context, and the wide range of policy responses, in the CWSRF program. These cases also illustrate that the underlying assumptions of Reagan federalism, while viable in some states, were clearly not viable in others. Of the four states discussed here, California presents itself as the state that was most clearly prepared to assume responsibility for the CWSRF program. It should be noted that Reagan was a long-time resident of California and had served as the state's governor from 1967 to 1975. It is reasonable to assume that his view of state capacity, and thus that of his presidential administration, was shaped in large part by his eight years as governor of one of the largest and wealthiest states in the nation.

This is not to suggest that the other three states discussed here were unprepared; rather, the relative capacity of the states differs. New York and Georgia had a great deal of experience with state-run and state-funded water quality programs, and that experience proved invaluable when implementation of the CWSRF began. Both states were able to employ existing agencies, personnel, and financial resources to implement their programs. On the other hand, Alabama lacked both financial resources and expertise to implement their program. Indeed, the initial lack of action on the part of the state to develop a program opened the door for the actions of the private bond banker to approach the legislature and propose a program structure. The state administrative and political apparatus was, for all intents and purposes, left on the sidelines and had effectively no input into the process. The long-term consequences are real and measurable: The requirements of aggressive leveraging mean that Alabama underserves small and financially at-risk communities, two of the three classes of communities highlighted in the WQA.

The program variability across these four states reflects the unique conditions in each state. New York must address large-scale, expensive needs in the metropolitan New York area, and smaller-scale needs in the rest of the state. The structure of their program specifically allows the state to meet those needs; such a program would not be viable in the two southern states. While California has several large coastal cities (including Los Angeles, with a population on par with that of New York City), California's cities are generally more self-sufficient and able to meet their needs without huge infusions of state assistance. As a densely populated, industrial state, New York has decades of environmental degradation to address, as evidenced by its huge water quality needs. In this sense, the states do serve as policy laboratories (Dror 1968), and the CWSRF program gives states the flexibility to tailor their programs to meet their unique circumstances.

9

The Distributional Impacts of the CWSRF

A National Analysis

The analyses in the preceding chapters highlighted the differences between states in terms of early design and implementation choices in the Clean Water State Revolving Fund (CWSRF) program. Taken collectively, these decisions shaped the implementation structures for state CWSRF programs, and set the course for the long-term operation of state programs. These decisions were not without consequence; they determined the degree to which states have been able to meet the intent of the Water Quality Act (WQA) and, ultimately, the ability of states to meet water quality needs.

This chapter examines the distribution patterns of CWSRF resources in states. By employing the model developed in Chapter 6, we seek to establish the determinants of state choice, and the degree to which state choice determines program outcomes. Our elements of state choice – primary institutional location of the CWSRF program, the number of agencies involved in implementation, the decision to leverage, the total number of actors in the CWSRF structure, initial sufficiency of state resources for the environmental and financial program elements, the number of years since a state assumed primacy in water quality, the number of National Pollution Discharge Elimination System (NPDES) authorizations held by a state, and a state's use of private sector actors – are arrayed against a set of demographic, political, and environmental control variables to explain differences in the distribution of CWSRF program resources (Table 9.1). The distribution of program resources is measured by examining resources allocated by states for significant environmental need, resources to small communities, resources to hardship communities, and resources to address nonpoint pollution. It is worth noting again the limitations of these analyses based on the lack of a suitable outcome measure for the program. The lack of a comparable and reliable measure of water quality across states limits the ability to tie state choices and actions to long-term results. In place of long-term outcomes, these analyses examine program outputs in terms of the dollar amounts and the number of loans made to the applicant categories of interest.

Table 9.1 *A model of state choice*

Category		Variable
Dependent variables		Loan dollars for environmental need; small communities; hardship communities; and nonpoint projects
		Percentage of loans for environmental need; small communities; hardship communities; and nonpoint projects
Independent variables		Program location
		Leverage status
		Number of agencies in CWSRF structure
		Initial state sufficiency of resources (environmental and financial)
		Primacy year
		NPDES authorizations
		Privatization score
Control variables	Demographic	State population density
		Median income
	Political	Government ideology score
		Legislative professionalism score
		Political culture
		Split government
		Democratic/Republican Trifecta
		State centralization score
		Federal region
	Environmental	State WQ needs, Category I and Category II
		NPS activism score (nonpoint only)

This chapter begins with a brief discussion of the methods employed in these analyses, and the sources of data used in the model. We then move to a series of analyses that examine program resource distribution at the national level. These national-level analyses serve both to set the stage for the state-level analyses, and to provide a baseline for overall program performance. The bulk of the chapter focuses on a series of regression models that test the model by looking at the distribution of program resources across the four recipient categories of interest. The chapter concludes with a summary of the major conclusions drawn from the analyses.

The primary source of CWSRF program data is the National Information Management System (NIMS) maintained by the US Environmental Protection Agency (EPA) to track CWSRF program performance. The NIMS reports consist of annual data across a set of more than 300 data points for each fiscal year. Several of the data points are calculated from data reported by states, but this study relies on raw data reported by states, as well as appropriation figures added annually by EPA. The current NIMS data, as of this writing, cover the period from FY 1988 through FY 2019; however, because of ongoing auditing and updating of data, we only include data through FY 2016.[1] Other variables in the dataset are drawn from a variety of additional sources; Appendix A provides a comprehensive list of data sources and definitions.

The national trend analyses presented in the first part of this chapter rely primarily on simple descriptive statistics and visual representations of relationships using two-axis graphs. The models in the latter part of the chapter utilize ordinary least squares (OLS) regression techniques estimated using Stata version 16.[2] Descriptive statistics and correlation tables may be found in Appendix B. We also present a series of figures to represent graphically both the trends in the data, as well as to more fully examine the marginal effects of several variables of interest.

National Trends in CWSRF Resource Distribution, 1988–2016

Before we address the implications of state choice in the CWSRF program, it is useful to cast a glance at the performance of the CWSRF program taken *in toto*, over the first thirty years of the program. This analysis provides a baseline for national program performance, and allows us to address two central questions of this book. First, has the CWSRF been successful as a means to meet water quality needs in the United States? Second, as a lasting expression of Reagan federalism, has the program lived up to its expectations of greater state responsibility? Were states prepared to assume responsibility for the CWSRF program, and if not, did that fact alter program outcomes? The lack of a clear set of outcome measures makes the former question more difficult to answer definitively, but there are several indicators that provide clues to the answer.

As detailed in Chapter 4, one of the motivations for the inclusion of the CWSRF in the WQA, and the phasing out of the construction grants program, was that states were simply replacing state effort with federal dollars. Instead of supplementing state funding efforts, the federal share of water quality dollars had outstripped total state effort to a significant degree. The WQA initially provided eight years of capitalization grants, at which time states would, in theory, be self-sufficient to meet their water quality needs. States were expected to provide the additional resources needed to meet their needs, either through additional state appropriation or, more likely, through the use of leveraging. Figure 9.1 plots total federal water quality funding (both Title II and Title VI) against total state funding (with and without leveraging). Even with leveraging, it took roughly five years for state funding to grow to a level higher than federal funding. The effect of leveraging is significant, particularly after about 2001.[3] It is useful to keep in mind, though, that fewer than half of all states have engaged in leveraging over the life of the program, which means that total state funding is driven by a few states. Absent leveraging, state funding has been, for the most part, anemic.[4]

In terms of cumulative spending on water quality, a similar picture emerges, but better demonstrates the effects of leveraging (Figure 9.2). Without leveraging, the

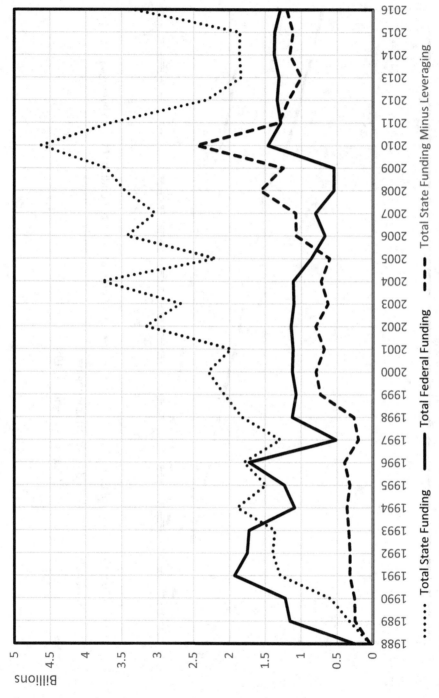

Figure 9.1 Comparison of total federal versus total state CWSRF spending, 1988–2016

.... Total State Funding   —— Total Federal Funding   ━ ━ Total State Funding Minus Leveraging

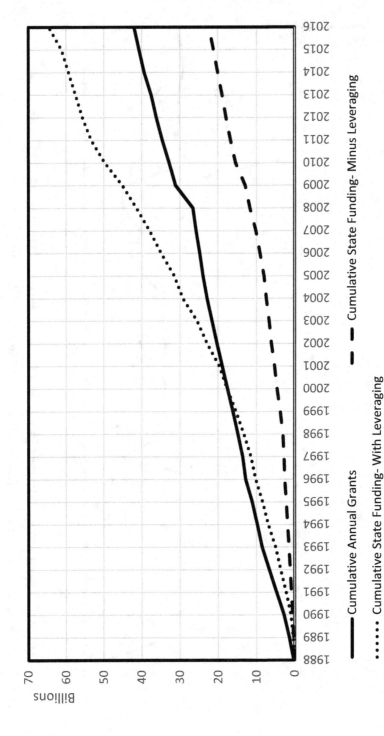

Figure 9.2 Cumulative CWSRF federal and state spending, 1988–2016

Cumulative Annual Grants

Cumulative State Funding- Minus Leveraging

Cumulative State Funding- With Leveraging

Table 9.2 *Comparative size of state efforts and CWSRF program*

Program	Number of Loans (Grants)	Total Dollar Amount	Average Dollars per Loan/Grant
CWSRF resources*	38,457	$91,893,655,072	$2,389,517
State loan programs	5,342	$9,335,372,634	$1,747,543
State grant programs	20,393	$10,178,962,815	$499,140
Total state effort	25,735	$19,514,335,449	$758,280

*Includes federal capitalization grants, ARRA grants, state match, excess match, and leveraged funds.

cumulative federal funding effort is nearly twice that of all states combined. Leveraging has increased total state effort to about threefold that of the total federal effort, but it is again worth noting that a majority of states have not engaged regularly in leveraging.[5] This would suggest that the goal of state self-sufficiency for water quality funding has yet to be met; indeed, without leveraging, the percentage gap between state and local effort reverts to a pattern similar to that seen between 1974 and 1983.[6] The trends since 1994 are also due to beyond-authorization appropriations made by Congress, along with the required state match. We might speculate that, absent the additional federal funding, state funding effort would likely have been significantly less after 1994.

A number of states also operate state-funded water quality programs outside the CWSRF structure. These programs may be loan or grant programs; a few states operate both kinds of state programs. The total size of these programs pales in comparison to the CWSRF program; Table 9.2 indicates the overall program outputs for the CWSRF program from 1988 to 2016. Notable is the average value of the loan/grant for state-funded programs as compared to the CWSRF; state assistance packages are about a third the size of CWSRF assistance packages.

In regard to the ability of the CWSRF program to meet state water quality needs, the evidence suggests that water quality needs continue to grow more quickly than CWSRF resources. Figure 9.3 plots water quality needs from the quadrennial Needs Surveys against total cumulative water quality spending (including federal capitalization grants, state match, leveraged funds, and state programs). The trend line makes clear there remains a growing gap between water quality spending and state water quality needs. While the CWSRF program has certainly contributed to thousands of projects that have addressed water quality, a combination of aging infrastructure and a growing population means that needs continue to increase more quickly than resources.

An alternative means to consider the ability of the CWSRF to meet national water quality needs is to focus specifically on the most significant environmental needs. These needs are listed by the EPA as Category I (secondary treatment) and Category II (advanced treatment) needs.[7] In this case a gap between needs and funding still exists (see Figure 9.4), but the gap is not widening in the same manner

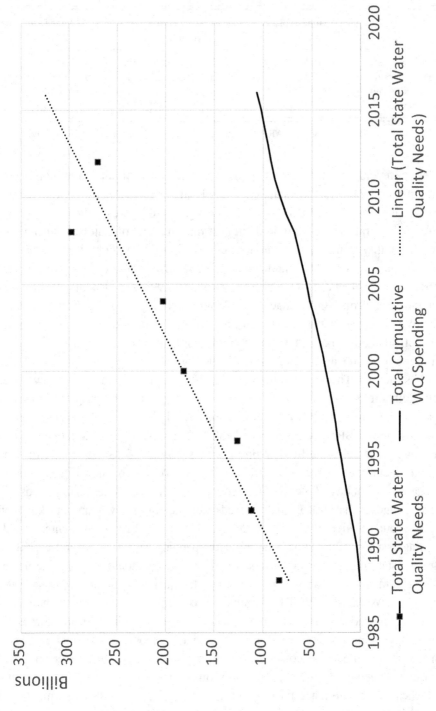

Figure 9.3 Comparison of total water quality funding effort to water quality needs

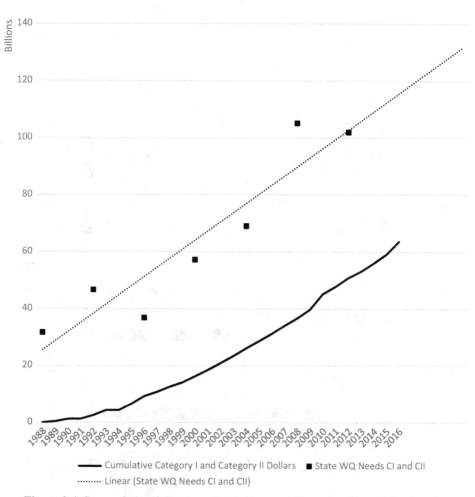

Figure 9.4 Comparison of Category I and Category II needs and total funding for Category I and Category II projects

as in the previous figure. Because of the "first use" requirement in Section 602, coupled with the visibility and political value of large, point-source infrastructure projects, these projects have received the bulk of CWSRF resources over the years.

When we examine needs over time, we see a mixed pattern in terms of progress toward meeting state needs.[8] Figure 9.5 compares needs by category for 1988, 2000, and 2012 (in 2012 dollars).[9] Although Category I (secondary treatment) and Category IV (new conveyance systems) needs have declined over the years, needs for advanced treatment (Category II), conveyance system repair (Category III), and combined sewer overflow (CSO) correction (Category V) have all increased over time. Stormwater management needs have also increased, although data for this need were not collected until 1992. Likewise, Category X (recycled water) needs were not collected prior to 2004.[10]

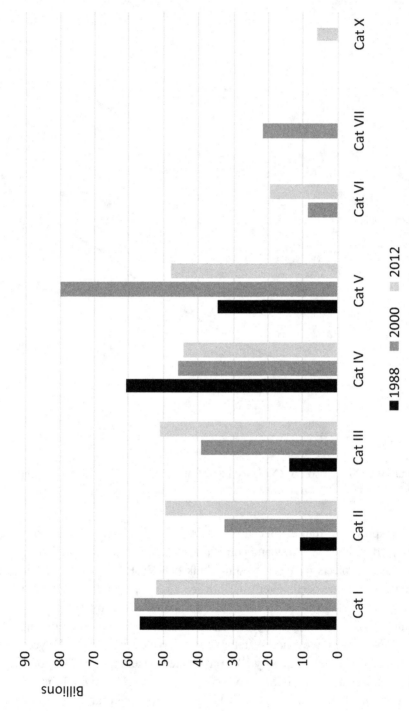

Figure 9.5 Comparison of wastewater needs by category, 1988, 2000, and 2012, in 2012 dollars

The quadrennial needs assessments do not report needs by potential recipient community type (e.g., small communities[11] or hardship communities), but rather only by need category. Likewise, there is no separate reporting (or a needs category) for nonpoint pollution. Still, the 2012 Needs Assessment (EPA 2016, 22) provides some clues to the needs of small communities, as well as rural communities. The report notes that, in 2012, about 12 percent of all needs (or about $32.9 billion) were for communities with populations less than 10,000. Similarly, rural needs were estimated at $67.8 billion, or about 25 percent of the total. The report also notes that about 80 percent of all centralized wastewater facilities serve small communities, which account for about 7 percent of the nation's population (EPA 2016, 23).

The EPA does not require states to report nonpoint-pollution needs, and there is no reasonable methodology to determine nonpoint needs. However, the National Information Management System does report cumulative dollars for nonpoint projects. As a proxy for needs, Figure 9.6 plots the cumulative CWSRF dollars for nonpoint projects against the cumulative grant dollars awarded to states under Section 319[12] of the WQA. Because states are required to have plans in place

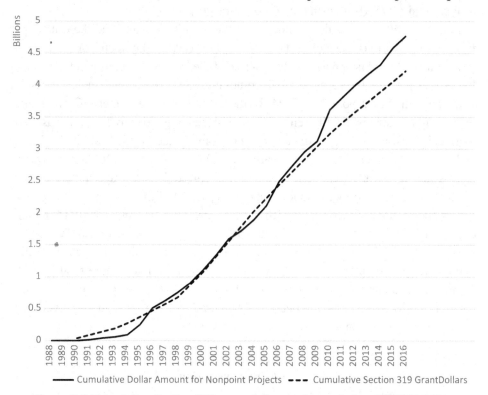

Figure 9.6 Cumulative Section 319 grant dollars and cumulative CWSRF dollars for nonpoint projects, 1988–2016

Table 9.3 *Average dollars per loan, 1988–2016*

Category	Number of Loans	Total Dollars	Dollars per loan
Cat I and Cat II needs	**13,791**	**$63,601,375,800**	**$4,611,803**
Pop <10k	**25,311**	**$26,576,910,781**	**$1,050,014**
Pop 10k–100k	9,577	$41,374,853,530	$4,320,231
Pop >100k	3,569	$50,779,147,425	$14,227,836
Hardship communities	**3,715**	**$9,211,099,423**	**$2,479,435**
Nonpoint projects	**16,717**	**$4,764,571,597**	**$285,014**
Native American tribes	16	$74,421,414	$4,651,338
Other*	310	$323,813,198	$1,044,559

*Includes estuary assistance, planning and assessment, and desalination.
Categories in bold indicate dependent variable categories.

under Section 319, we can estimate the degree to which those plans have spurred nonpoint-project investment in the CWSRF program. No state made a CWSRF loan for nonpoint uses until 1991, and several states have never made loans for nonpoint projects.[13] Still, it appears that, overall, the Section 319 program has led to increased CWSRF investment in nonpoint projects.

Finally, Table 9.3 lists the number of loans, the total dollars, and the dollars per loan for the various categories tracked in the NIMS. The focus on significant environmental need is apparent both in terms of the number of loans and the average loan size. Likewise, the focus on small communities is also evident, and is consistent with the estimates reported by EPA in the 2012 needs assessment. Of particular interest in this table is the difference in size of the average loan to different classes of loan recipients. For example, the average nonpoint project receives about $285,000, and states collectively have made nearly 17,000 of these loans. On the other hand, large-scale wastewater treatment works in the largest cities average more than $14 million per loan, yet only account for about 20 percent of the number of loans made for nonpoint uses. Since nonpoint-pollution projects are less likely to require expensive infrastructure, it is possible to spread a smaller amount of money across a larger number of recipients than is the case for large-scale projects. Likewise, smaller systems receive smaller amounts of assistance, but states make more total loans to communities under 100,000 population. Finally, virtually no CWSRF resources have been directed to serve the needs of Native American tribes, and relatively few resources have been directed to other allowable uses.

A Note on Leveraging

The effects of leveraging on the abilities of states to meet different categories of needs can be detected by a comparison of means between leveraged states and non-leveraged states. To begin with, EPA guidance (1988a) suggests that states with greater water quality needs should leverage their capitalization grants to raise additional dollars to make loans to applicant communities. Additionally, states that leverage will be less likely to appropriate additional dollars (beyond the 20 percent match) to the CWSRF. Table 9.4 indicates no significant difference in means between the water quality needs of leveraged and non-leveraged states. As expected, leveraged states have, on average, about $2 billion more in needs than do non-leveraged states, but the relative difference is small. On the other hand, the mean difference between leveraged and non-leveraged states in terms of additional state dollars is significant; on average, non-leveraged states contribute about $88,000 more than leveraged states. It is worth noting, however, that few states contribute any additional state dollars, so the magnitude of those contributions over the life of the CWSRF program is miniscule.

Another method to examine the relationship between leveraging and needs is with a scatter plot of water quality needs for Category I and Category II projects and leveraging amount (Figure 9.7). The graph illustrates the clusters that exist

Table 9.4 *Difference of means test for leveraged and non-leveraged states, water quality needs and additional state contribution*

Water quality needs, by leveraging status

Group	Obs.	Mean	Std. Err.	Std. Dev.	[95% Conf.	Interval]
Non-leveraged	876	1,577.464	84.22967	2,492.971	1,412.148	1,742.78
Leveraged	574	1,764.937	130.6539	3,130.246	1,508.318	2,021.556
Combined	1,450	1,651.677	72.56919	2,763.352	1,509.326	1,794.029
diff		−187.4726	148.3621		−478.5002	103.555

diff = mean(0) − mean(1) t = −1.2636
Ho: diff = 0 degrees of freedom = 1,448
Ha: diff < 0 Ha: diff != 0 Ha: diff > 0
Pr(T < t) = 0.1033 Pr(|T| > |t|) = 0.2066 Pr(T > t) = 0.8967

Additional state contribution, by leveraging status

Group	Obs.	Mean	Std. Err.	Std. Dev.	[95% Conf.	Interval]
Non-leveraged	876	122,257.6	28,181.56	834,097.9	66,946.29	177,569
Leveraged	574	36,590.95	8,461.841	202,731.3	19,970.94	53,210.96
combined	1,450	88,345.44	17,382.7	661,913.5	54,247.49	122,443.4
diff		85,666.69	35,485.83		16,057.56	155,275.8

diff = mean(0) − mean(1) t = 2.4141
Ho: diff = 0 degrees of freedom = 1,448
Ha: diff < 0 Ha: diff != 0 Ha: diff > 0
Pr(T < t) = 0.9921 Pr(|T| > |t|) = 0.0159 Pr(T > t) = 0.0079

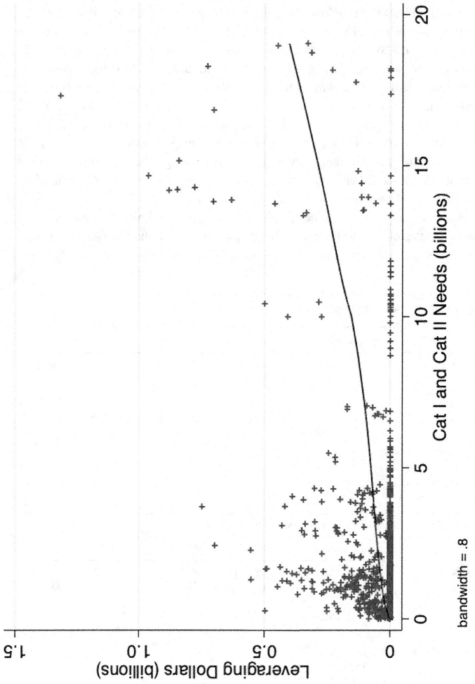

bandwidth = .8

Figure 9.7 Scatter plot of leveraging dollars and Category I and Category II needs

around different levels of needs. Leveraging seems to be most common when needs are less than $50 billion. While the LOWESS smoothed line trends upward as needs grow, the line is influenced by a relatively small number of observations with needs in excess of about $140 billion and who leverage more than about $300 million. This group of observations is from states such as New York, Massachusetts, and Pennsylvania, all of whom have leveraged aggressively and have significant water quality needs. For most observations, however, states that leverage tend to have relatively low water quality needs.

When we examine the effects of leveraging on resources directed to different kinds of applicant community, a different picture emerges. The ability of leveraging to increase the resources available in the CWSRF is considerable; moreover, the effect of leveraging is positive on the distribution of resources to all four of our community types of interest (Table 9.5). Not surprisingly, the mean difference is largest for communities with significant environmental need; the large dollar value of those loans (see Table 9.3) requires a substantial commitment of resources. Likewise, the relatively low value of nonpoint projects supports the comparatively smaller difference in means between leveraged and non-leveraged programs. Leveraging is thus not directly tied to the magnitude of state water quality needs, but clearly does result in more resources available for all categories of applicant.

Summary

Our analyses of national trends suggest that states have yet to reach self-sufficiency for water quality funding. The promise of the Water Quality Act and its then-revolutionary Clean Water State Revolving Fund was to turn responsibility for water quality infrastructure over to states, and thus remove the need for federal spending for water quality. Both of these outcomes have so far failed to be realized. Between 1995 and 2016 the national government invested roughly $32 billion beyond the Act's authorization,[14] and there is little indication that federal spending for water quality will terminate soon.[15] At the same time, state funding has not met expectations, in large part because of the decision of many states to eschew regular leveraging. The few states that have leveraged have brought total state effort above total federal effort; however, state needs for water quality funding continue to exceed available combined resources. These analyses also suggest that the focus of the CWSRF program has been to fund environmental need above other needs. The bulk of CWSRF resources have been allocated to meet Category I and Category II needs. Due to ongoing appropriations, the "first use" requirement applies to each capitalization grant, which ensures that the bulk of new resources will go to meet significant environmental needs. Small

Table 9.5 *Difference of means test for applicant community types, by leveraging status*

Environmental Need

Group	Obs.	Mean	Std. Err.	Std. Dev.
Non-leveraged	876	4.71E+07	2,532,944	7.50E+07
Leveraged	574	7.03E+07	4,993,858	1.20E+08
Combined	1,450	5.63E+07	2,516,574	9.58E+07
diff		−2.32E+07	5,111,636	

diff = mean(0) − mean(1) t = −4.5330
Ho: diff = 0 degrees of freedom = 1,448
Ha: diff < 0 Ha: diff != 0 Ha: diff > 0
Pr(T < t) = 0.0000 Pr(|T| > |t|) = 0.0000 Pr(T > t) = 1.0000

Hardship Communities

Group	Obs.	Mean	Std. Err.	Std. Dev.
Non-leveraged	876	6,379,030	664,654.8	1.97E+07
Leveraged	574	1.14E+07	1,357,362	3.25E+07
Combined	1,450	8,358,618	673,579.1	2.56E+07
diff		−5,000,701	1,371,558	

diff = mean(0) − mean(1) t = −3.6460
Ho: diff = 0 degrees of freedom = 1,448
Ha: diff < 0 Ha: diff != 0 Ha: diff > 0
Pr(T < t) = 0.0001 Pr(|T| > |t|) = 0.0003 Pr(T > t) = 0.9999

Small Communities

Group	Obs.	Mean	Std. Err.	Std. Dev.
Non-leveraged	876	1.92E+07	906,369.5	2.68E+07
Leveraged	574	2.78E+07	1,333,807	3.20E+07
Combined	1,450	2.26E+07	768,465.4	2.93E+07
diff		−8,648,574	1,555,416	

diff = mean(0) − mean(1) t = −5.5603
Ho: diff = 0 degrees of freedom = 1,448
Ha: diff < 0 Ha: diff != 0 Ha: diff > 0
Pr(T < t) = 0.0000 Pr(|T| > |t|) = 0.0000 Pr(T > t) = 1.0000

Nonpoint Source Projects

Group	Obs.	Mean	Std. Err.	Std. Dev.
Non-leveraged	876	3,272,298	414,763.9	1.23E+07
Leveraged	574	4,395,873	795,284.4	1.91E+07
Combined	1,450	3,717,079	402,463.3	1.53E+07
diff		−1,123,574	822,728.7	

diff = mean(0) − mean(1) t = −1.3657
Ho: diff = 0 degrees of freedom = 1,448
Ha: diff < 0 Ha: diff != 0 Ha: diff > 0
Pr(T < t) = 0.0861 Pr(|T| > |t|) = 0.1723 Pr(T > t) = 0.9139

communities have also received attention from CWSRF funds, although hardship communities have fared less well. States were hesitant to fund nonpoint projects in the early years of the program but, collectively, states have been much more willing to direct CWSRF resources for nonpoint pollution in the last twenty-five years. While generally low in dollar value, the number of loans indicates a commitment on the part of some states to address nonpoint needs.

State Choice and National Loan Distribution Patterns

We now turn our attention to explanations of state behavior in the CWSRF program. By applying the model developed in Chapter 6, we can explore the parameters of state choices in the CWSRF program and determine how those choices impact the ability of individual states to meet the intent of Congress in terms of the distribution of CWSRF resources. To this end, we will apply the model of state choice to four categories of loan recipients: loans for significant environmental need, loans to small communities, loans to hardship communities, and loans for nonpoint pollution. Each dependent variable will be measured in two ways: the dollar value of loans, and the percent of total loans made for that purpose. It is also important to note that the categories in the NIMS data are often not exclusive, particularly in terms of the classes of applicants. In other words, a single loan could be executed with a small community with Category I needs, and that loan would appear in both categories in NIMS. This is not necessarily problematic in analytical terms, since it makes sense that a single loan could meet more than one purpose. Indeed, it would be possible for a single loan to meet up to three uses: environmental need, small community, and hardship community; or small community, hardship community, and nonpoint source.[16]

How Well Have States Addressed Significant Environmental Need?

The first category of recipients to consider is communities with significant environmental need. The WQA defines this category as communities with Category I (secondary treatment) or Category II (advanced treatment) needs. Communities with these needs are likely to be in danger of NPDES violations, and these communities are also the intended target for the "first use" requirement of the law. These needs are also explicitly point-source pollution needs.

As noted previously, the focus of the WQA is on point-source pollution, and the history of water quality policy has largely emphasized large-scale water quality infrastructure to address point-source pollution issues. These are also the most expensive issues to address, so it is no surprise that the bulk of CWSRF resources are allocated for this purpose. While a model using the percent of loans for

Table 9.6 *OLS regressions for environmental need: Dollars and percent loans*

	Dollars for Environmental Need		Percent Loans for Environmental Need	
	Coefficient	Robust Std. Err.	Coefficient	Robust Std. Err.
Independent variables				
Institutional location				
Health	5.409986	7.144182	−17.54777***	6.834597
Financial	−3.794103	10.74676	−14.74325***	5.069629
Other	5.498863	9.96824	1.093036	27.01093
Leveraging $	0.3347567***	0.054594	0.0164661***	0.0047644
Total actors	−13.47995	20.62415	−18.8011*	10.7604
Initial sufficiency – Env.	18.43887	40.00793	46.27109**	20.30747
Initial sufficiency – Fin.	−68.97583	53.79585	−56.65453**	24.83523
Years of primacy	1.118533***	0.2750785	−0.0504283	0.1265287
NPDES authorizations	14.05334	12.11903	8.304419	7.137355
Privatization score	−0.6810619	5.889186	9.066959**	3.613016
Control variables				
Population density	0.1902815	0.1433282	−0.0209561	0.0526463
Median income	0.0002584*	0.0001522	0.0000937	0.0000628
Government ideology	−0.2142528	0.2126684	−0.1240265	0.0838973
Centralization	−0.8043823	1.005214	−0.5513436	0.5970946
Moralistic	8.732598	10.33195	33.62252***	7.17027
Individualistic	38.25196	74.95488	70.59568	43.81572
GOP Trifecta	−13.22351***	4.126129	−3.647806*	2.053214
Dem Trifecta	7.427636	5.487255	5.811214***	1.6389
Professionalism	−127.2095*	67.61138	−41.56818***	15.68112
WQ needs	0.0050611	0.004136	−0.0005674	0.0005159
Constant	65.08753	99.32848	87.43764**	40.00075
N	1,192		1,192	
Adjusted R^2	0.7065		0.315	

*p > .1; **p > .05; ***p > .01
Note: State fixed effects estimated but not presented.

environmental need is presented, the more conceptually interesting model uses dollars for environmental need. The first model (see Table 9.6) clearly illustrates the importance of leveraging on the distribution of CWSRF dollars for significant environmental need. States leverage to increase the dollars available for loans, and it follows that those resources should be directed toward the most expensive needs.

This model also suggests that the hypothesized relationships between environmental need and other state implementation choices are generally supported. As the number of actors in the CWSRF structure increases, fewer dollars are allocated for environmental need, which is consistent with the hypothesized relationship, although the variable fails to reach statistical significance.

The privatization variable is negative for dollars for environmental need, and positive for the percent loans model, although it is only statistically significant in the model for percent loans. And, as environmental need generates the largest loan size, the positive effect of greater privatization becomes more evident.

There are also some less intuitive results in these models. For example, institutional location does not matter much in terms of dollars for environmental need, although environmental agencies (the reference category) generate a larger percentage of loans for environmental need than either health or financial agencies. Perhaps more surprising is the negative relationship between dollars for environmental need and initial sufficiency of financial personnel. This suggests that states who were initially lacking in financial expertise have corrected those deficiencies over time. This issue will be explored in more detail later in this chapter. The relationship between environmental need and the number of NPDES authorizations held by a state is positive but not significant. The variable for years of primacy is both positive and significant, suggesting that states who achieve primacy earlier in time are more likely to direct more resources to environmental need. A likely explanation for this is that states who sought primacy early have a stronger commitment to both water quality and the ability to retain state control of the regulatory and enforcement regime. This explanation would be consistent with Davis and Lester (1989), who noted that states differed in their commitment to environmental protection.[17]

The control variables present mixed results. All of the variables are in the hypothesized direction, even if the coefficients fail to reach statistical significance. States in which the Republican Party control both chambers of the legislature and the governor's mansion tend to spend about $13 million per year less than states with split government, which provides some support for an explanation focused on party competition. Of particular note is the strong difference in both dollars and percentage of loans for environmental need between traditionalistic states (the reference category) and other states. Overall, the model for environmental dollars explains about 70 percent of the variance in the dependent variable.

Figure 9.8 illustrates the marginal effects of leveraging on resources dedicated to environmental need. The point of leveraging is to generate more dollars for loans; environmental need loans are (on average) the most expensive projects, and thus the need for additional resources to meet these is most acute compared to other uses. The effect of leveraging a million dollars to $36 million results in roughly 25 percent more dollars allocated for significant environmental need.

How Well Have States Addressed Small Community Needs?

The EPA reports that states allocate about 80 percent of their CWSRF resources to small communities (EPA 2016). Our regression model for dollars allocated for small communities (see Table 9.7) indicates a positive effect of leveraging; the real

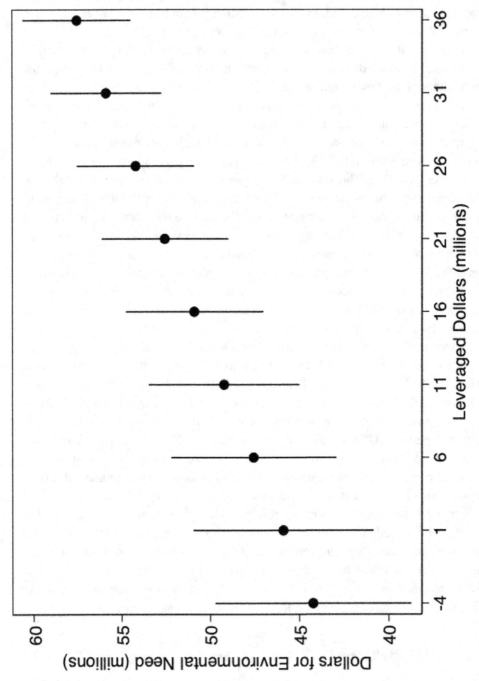

Figure 9.8 Marginal effects of leveraging on dollars for environmental need

Table 9.7 *OLS regressions for small communities: Dollars and percent loans*

	Dollars for Small Communities		Percent Loans for Small Communities	
	Coefficient	Robust Std. Err.	Coefficient	Robust Std. Err.
Independent variables				
Institutional location				
Health	2.983687	3.592551	−8.443973	6.744843
Financial	14.61869***	4.72709	2.082235	4.798456
Other	8.641546**	3.855258	25.92542***	7.297919
Leveraging $	0.0246766*	0.0126422	−0.0009612	0.0059508
Total actors	15.84984	10.23275	5.394724	13.09144
Initial sufficiency – Env.	−35.47094*	18.64076	−15.64132	24.88693
Initial sufficiency – Fin.	2.186988	21.57682	19.00806	28.62232
Years of primacy	0.2315401**	0.0998694	0.1953865	0.135424
NPDES authorizations	−2.084901	5.764634	2.538875	8.498343
Privatization score	−8.342931**	3.357675	−6.922799*	4.119755
Control variables				
Population density	0.0838812**	0.0430793	0.1615491**	0.0765062
Median income	−0.000044	0.0000623	−0.0000686	0.0000713
Government ideology	−0.204731***	0.0761434	−0.0793576	0.1017705
Centralization	1.067925**	0.4944776	1.050548*	0.614004
Moralistic	−30.64702***	8.518171	−5.007884	8.440008
Individualistic	−66.02039	40.74666	−13.38925	50.70351
GOP Trifecta	−3.225985	2.435154	2.15519	2.157688
Dem Trifecta	2.536661	1.670458	3.699**	1.819095
Professionalism	−30.36983*	17.44505	−12.27284	17.81125
WQ needs	0.0025568***	0.0009055	0.0003545	0.0006006
Constant	−27.19487	33.52658	−35.32215	48.39252
N	1,192		1,192	
Adjusted R^2	0.475		0.326	

*p > .1; **p > .05; ***p > .01.

Note: State fixed effects estimated but not presented.

effect of leveraging is to make more dollars available for loans. Leveraging is not significant in terms of the percent of loans for small communities, but given the smaller value of these loans, this result is not surprising. This model also indicates that, for small communities, financial and other lead agencies are more likely to allocate dollars to small communities than are environmental agencies. We see a negative, but statistically insignificant, relationship between dollars for small communities and NPDES authorizations, but a positive relationship between the dependent variable and years of primacy. Although the variables fail to reach statistical significance, states with a greater role for private actors tend to direct fewer resources to small communities; given EPA's (2016) suggestion that some 80 percent of all CWSRF resources are directed to small communities, the decrease in the amount of dollars (and percent loans) in the CWSRF suggests that the private sector may focus more on large-scale (high-dollar) projects.

Of more interest is the positive relationship between the dependent variables and population density. This suggests that more densely populated states direct more dollars to small communities, which is somewhat counterintuitive. The variable for water quality needs is significant for the dollar amounts directed to small communities, providing additional support for the emphasis states place on small community needs. However, the coefficient is relatively small, which suggests that water quality needs, while not unimportant, has a minor impact on the CWSRF resources directed by states to small communities. Overall, the model for dollars for small communities explains about 47 percent of the variance, and about 32 percent of the variance in the percentage loan dependent variable.

When we examine the marginal effects of the total number of actors in the CWSRF structure, we see the effect on the dollars allocated to small communities (Figure 9.9). As the number of actors in the CWSRF structure increases, the dollars allocated to small communities tends to decrease. There is a slight increase for states with seven and eight actors, but there are only two states with seven actors, and three states with eight actors (a total of 138 observations out of 1,450 total observations). The trend does suggest that, for dollars to small communities, the organizational and decisional complexities that accompany multiple decision points in implementation suggested by Montjoy and O'Toole (1984) and Pressman and Wildavsky (1973) is supported.

How Well Have States Addressed Hardship Community Needs?

Of the three kinds of community addressed in this study, the water quality needs of hardship communities are perhaps the most challenging needs to meet. Hardship communities tend to be poor, and less capable of absorbing large increases in user fees to repay a loan. As reported earlier in this chapter, loans to hardship communities tend to be about twice as big, on average, as loans to small communities, but only about half the amount of the average loan for environmental need. In addition, the total number of loans made to hardship communities is significantly smaller: States have made only about 27 percent of the number of loans made for environmental need; about 22 percent of the number of loans made for nonpoint projects, and only about 15 percent as many loans as made for nonpoint projects small communities. The nature of a policy instrument that relies on loans (as opposed to grants) makes it more difficult for states to meet the needs of communities who will struggle to pay back a loan.

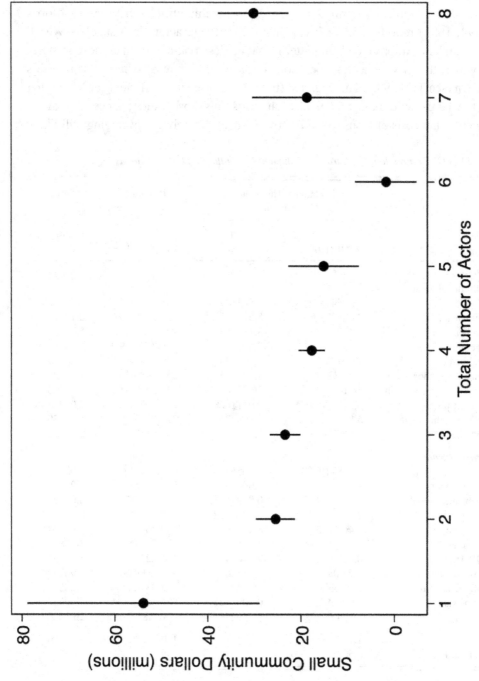

Figure 9.9 Marginal effects of the total number of actors in the CWSRF structure on dollars to small communities

Perhaps indicative of this tension is the effect of private sector activity on dollars to hardship communities; a one-unit change in the privatization score results in about $11 million fewer dollars to hardship communities each year. As Morris (1994, 1997) reports, greater involvement by private actors in a state's CWSRF program tends to produce loan patterns that are less risk-averse than those in states in which the private sector does not play a role. Our results here illustrate this pattern (see Table 9.8). The negative (but not significant) coefficient for leveraging in the hardship dollars model also supports this observation; since leveraging increases the costs of loans to the borrower, states that rely on leveraging will likely

Table 9.8 *OLS regressions for hardship communities: Dollars and percent loans*

	Dollars for Hardship Communities		Percent Loans for Hardship Communities	
	Coefficient	Robust Std. Err.	Coefficient	Robust Std. Err.
Independent variables				
Institutional location				
Health	12.14498***	2.874687	−0.9286216	5.173209
Financial	24.37025***	5.571549	8.243108***	2.134717
Other	7.47065**	3.133755	−2.194214	5.276947
Leveraging $	−0.0084664	0.0096283	−0.0078389**	0.0036499
Total actors	18.04824	12.08124	−20.33897**	8.143217
Initial sufficiency – Env.	−37.82282*	22.78605	55.88983***	15.68996
Initial sufficiency – Fin.	28.14929	26.33102	−39.89474**	17.54733
Years of primacy	0.2483001***	0.0811028	0.087023	0.0735735
NPDES authorizations	5.886012	5.868476	22.00878***	5.014395
Privatization score	−10.73435***	3.625579	1.636517	2.289224
Control variables				
Population density	0.1687107***	0.0656843	0.0635935*	0.0330321
Median income	−0.0000255	0.0000551	0.0000158	0.0000367
Government ideology	−0.1167455	0.0793228	−0.0744977	0.0532313
Centralization	2.107853***	0.4265332	1.080848***	0.3179841
Moralistic	−46.10638***	8.47206	5.598623	4.983671
Individualistic	−66.21931	43.75312	77.37022**	30.58073
GOP Trifecta	0.0759954	2.146342	1.206226	1.327326
Dem Trifecta	−2.608579**	1.255268	−0.4633704	0.9674745
Professionalism	18.06468	16.54255	13.03376	8.885702
WQ needs	0.0013873*	0.0007125	0.0008883***	0.0002836
Constant	−154.9982***	34.59256	−87.88528***	26.50864
N	1,192		1,192	
Adjusted R^2	0.507		0.49	

*p > .1; **p > .05; ***p > .01
Note: State fixed effects estimated but not presented.

make CWSRF resources even less attractive (or possible) for hardship communities. The strong negative coefficient for initial sufficiency of environmental resources in the dollar model is surprising, although initial sufficiency for both categories is both significant and in the expected direction for the percent loans model.

Also of particular interest is the effect of institutional location. In the model for dollars, environmental agencies are much less likely to fund hardship communities than are health, financial, or other agencies. This finding may be indicative of the tension between the environmental and financial components of the program. State environmental agencies tend to have, as one might expect, an environmental focus. Administrators in environmental agencies are likely to be caught between a desire to meet environmental needs and the requirement to maintain the CWSRF fund in perpetuity. It may be the case that, when faced with a marginal applicant in terms of creditworthiness, decision makers in environmental agencies take a more conservative approach to their loan decisions. This issue likely requires more in-depth exploration to determine why environmental agencies are less likely to fund hardship communities than CWSRF programs in other agencies.

Other findings from these models largely follow the patterns seen in the previous two models. The coefficients for years of primacy and the number of NPDES authorizations are positive, and years of primacy is significant for dollars for hardship communities although the coefficient for years of primacy is relatively small. A Democratic trifecta tends to reduce the dollars available to hardship communities by about $2.6 million per year. Water quality needs is significant in both models, but the coefficient is small. Overall, each of these models explains roughly half of the variance is its respective dependent variable.

Figure 9.10 illustrates the marginal effects of institutional location on dollars allocated by states to hardship communities. The findings here are somewhat counterintuitive; of particular note is the difference between environmental agencies and financial agencies. We might posit that an environmental agency is more concerned with environmental protection, and less focused on the financial stability of the applicant community. Likewise, financial professionals might be less willing to make loans that carry more inherent risk, such as loans to hardship communities. The opposite appears to be true. A different explanation might be found in two related arguments. First, environmental agencies are generally composed of environmental engineers rather than financial experts. This may lead to a relatively conservative outlook on the part of environmental agencies, and make them less likely to take risks with scarce resources. Likewise, states that leverage more aggressively (and thus have more dollars available for loans) tend to locate their programs in financial agencies. The additional dollars available, coupled with a more fine-tuned appreciation for the specific amount of risk inherent in a particular applicant, may lead financial agencies to direct more resources to hardship communities.

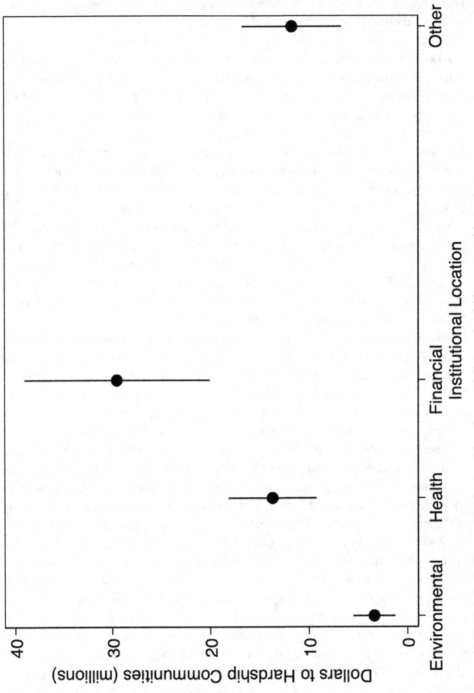

Figure 9.10 Marginal effects of institutional location on dollars to hardship communities

How Well Have States Addressed Nonpoint-Source Pollution Needs?

Most states are relative newcomers to the world of nonpoint pollution in the CWSRF program. Few states made any loans for nonpoint projects until about 2000, and several states (e.g., Alabama) have never made a loan for nonpoint uses. In recent years, however, states have made more resources available to meet nonpoint-pollution needs. Because of the relatively small average dollar value of these loans (about $285,000 per loan), the dollars spent on nonpoint projects is miniscule compared to other categories of applicants. For this reason, we will focus on the percentage of loans for nonpoint as our target for analysis (see Table 9.9). The regression results for nonpoint dollars are presented here for consistency.

Table 9.9 *OLS regressions for nonpoint-source projects: Dollars and percent loans*

	Dollars for Nonpoint Projects		Percent Loans for Nonpoint Projects	
	Coefficient	Robust Std. Err.	Coefficient	Robust Std. Err.
Independent variables				
Institutional location				
Health	3.368893***	1.145405	4.525475**	2.206024
Financial	9.85331***	3.386696	7.297998***	2.843528
Other	1.563033	1.109685	1.035463	2.534919
Leveraging $	0.0220383	0.0186858	0.0016257	0.004227
Total actors	17.64508	18.96494	−188.0531***	52.89665
Initial sufficiency – Env.	−34.74022	35.07777	345.9876***	98.42231
Initial sufficiency – Fin.	43.03882	50.71871	−558.6117***	144.3753
Years of primacy	−0.012339	0.05778	0.3954627***	0.0840496
Privatization score	−4.787657	4.532591	37.26333***	12.42236
Control variables				
Population density	0.0158532	0.0194338	0.1181787**	0.0602353
Median income	3.52E-06	0.0000221	8.02E-06	0.0000451
Government ideology	−0.0884411*	0.0470853	0.0738835	0.0707872
Centralization	0.2257041	0.1520813	2.880257***	0.383185
Moralistic	−17.53605	16.07781	132.38***	44.1442
Individualistic	−45.98953	59.63636	611.9169***	166.1133
GOP Trifecta	0.1830899	0.7813159	−2.293379	1.557456
Dem Trifecta	−0.8515143	0.6334424	−2.682487**	1.203989
Professionalism	−12.37958	8.637495	−19.48829**	9.090569
WQ needs	0.0001457	0.0007111	−0.0005363*	0.000319
NPS activism	1.054878	1.088012	−12.27365***	2.899841
Constant	−71.68067	75.99623	713.1658***	210.4261
N	1,192		1,192	
Adjusted R^2	0.338		0.600	

*p > .1; **p > .05; ***p > .01
Note: State fixed effects estimated but not presented.

The model for percent loans is relatively robust, explaining about 60 percent of the variance in the dependent variable. The coefficients for total number of actors, initial sufficiency (both categories), and years of primacy are all statistically significant, and consistent with our initial expectations in terms of the direction of the relationship. The effects of privatization in the CWSRF program are also apparent; states with a greater role for the private sector are more likely to direct resources for nonpoint projects. Leveraging is not significant; given the relatively low dollar requirements for nonpoint projects, leveraging provides little advantage for this use. The other independent variable of interest in this model is institutional location. Environmental agencies are less likely to direct CWSRF resources to nonpoint projects than other agency types. This suggests that environmental agencies may have an institutional bias toward traditional point-source pollution. Water quality offices in environmental agencies tend to be staffed with wastewater engineers, which may lead to a preference for projects that require large-scale engineering solutions. In addition, the regulatory function (the NPDES program) is typically located in an environmental agency. Since the NPDES program is explicitly focused on point-source pollution and nonpoint pollution lacks the same sort of formal regulatory regime, it is likely that the culture and practice of these agencies creates a symbiotic and mutually supportive relationship between the funding and regulatory functions, resulting in a narrower focus on traditional point-source pollution.

In terms of the control variables in this model, the coefficients for both Republican and Democratic trifecta are negative and significant, as is the coefficient for the professionalism measure. These models also employ a measure of nonpoint-source activism drawn from Hoornbeek (2011). As noted in Chapter 6, the expectation is that states with higher NPS activism scores will direct more CWSRF resources to nonpoint projects than states with lower activism scores. Surprisingly, the NPS activism measure reaches statistical significance, although the coefficient is opposite the hypothesized direction. Hoornbeek (2011) created this variable in the mid-2000s; it is likely that the change in focus on nonpoint pollution in recent years has changed.[18] States with Democratic control of state government are slightly less likely to direct loans to nonpoint projects, although the size of the effect is relatively small. On the other hand, states with a greater degree of centralization are slightly more likely to fund nonpoint projects.

The effects of political culture are particularly strong in these models; both moralistic and individualistic states are much more likely to direct a greater percentage of loans to nonpoint pollution. On one hand, we might expect moralistic states to have a greater interest in addressing nonpoint pollution. Traditionalistic states tend to be more rural, and have a larger agricultural presence,[19] but agricultural interests tend to be strongly opposed to nonpoint-pollution controls (see Houck 2002). Although there are many sources of nonpoint pollution, agricultural runoff is a significant contributor to nonpoint pollution. Unlike private homes in a neighborhood, farms are also business entities, and are more likely targets for loans than are individual citizens or property owners.

The marginal effects of institutional location on the percent of loans made for nonpoint projects can be seen in Figure 9.11. As noted earlier, the most likely explanation for the relatively low performance in this category by environmental agencies is an institutional bias for large-scale, complex infrastructure solutions for water pollution. The distinction between environmental agencies and health agencies seems to support this conclusion; while health agencies tend to direct fewer loans to nonpoint projects than do financial agencies, health agencies make about 5 percent more loans for nonpoint projects than environmental agencies.

Comparison across the Models

While the findings presented in this chapter generally support the model introduced in Chapter 6 (see Table 9.10), there are several findings that bear additional insight and theoretical development. For example, the mostly negative relationship between our measure of initial sufficiency for financial resources and the dependent variables suggests at least three possibilities. First, it is possible that the measure of initial sufficiency fails to capture the true state of initial sufficiency for financial personnel across the states. A second interpretation is that states that lacked resources at the outset were able to acquire needed resources over time, and that the cumulative effect of initial insufficiency is overwhelmed over the life of the program. Finally, a third possibility is that initial sufficiency of financial resources simply has no effect on state distribution patterns, and the model needs to be reconsidered. On the other hand, while not always significant, greater initial sufficiency of environmental personnel tends to enhance the distribution of loan resources to these categories of recipients.

To test the explanation that initial sufficiency of financial personnel becomes less important over time, the models for environmental need, small communities, and hardship communities were estimated again, each time dropping the most recent five years of data for each iteration. The same technique was applied to the percent loans for nonpoint pollution. The results indicate that the magnitude of the coefficient was reduced in each successive model, and by 1996 the coefficients became positive and significant. This suggests that states were able to overcome this initial insufficiency.

The findings for the total number of actors in the CWSRF structure are consistent with the expectations of the model, although the variable only reaches statistical significance for percent loans for nonpoint uses. The hypothesis suggested by the work of Montjoy and O'Toole (1984) and Pressman and Wildavsky (1973) is thus

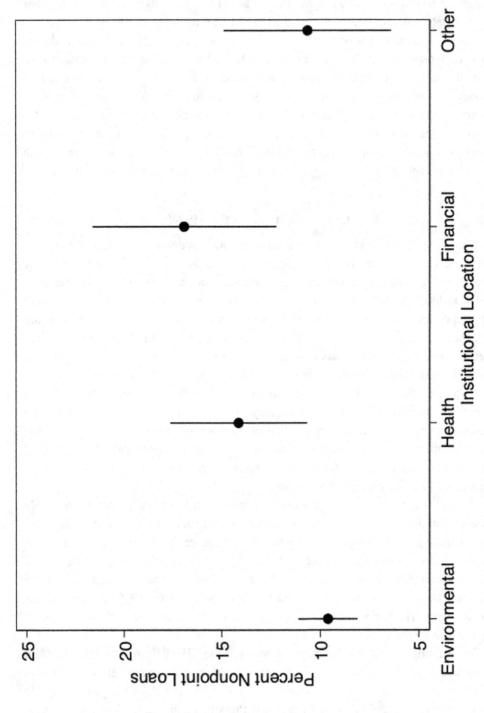

Figure 9.11 Marginal effects of institutional location on the percentage of loans for nonpoint projects

Table 9.10 *Overall findings for regression models*

Variable	Significant in Models?			
	Env. Need Dollars	Small Comm. Dollars	Hardship Comm. Dollars	Nonpoint Percent Loans
Independent variables				
Institutional location	N	Y	Y	Y
Leveraging	Y	Y	N	N
Total actors	N	N	N	N
Initial sufficiency – Env.	N	Y	Y*	N
Initial sufficiency – Fin.	N	N	N	N
Years of primacy	Y	Y	Y	Y
NPDES authorizations	N	N	N	Y
Privatization	N	Y	Y	Y
Control variables				
Population density	N	Y	Y	Y
Median income	Y	N	N	N
Government ideology	N	Y	N	N
Centralization	N	Y	Y	Y
Political culture	N	Y*	Y*	Y*
GOP Trifecta	Y	N	N	N
Dem Trifecta	N	N	Y*	Y*
Professionalism	Y	Y*	N	Y*
WQ needs	N	Y	Y	Y*
Nonpoint activism	–	–	–	N

*Variable was significant, but in the opposite direction from the hypothesized relationship.

supported for percent of nonpoint loans, but not for other dependent variables. A closer examination of the data reveals that the mean number of actors across all states is 3.6 (mode = 3), and only a handful of states now use a single agency for implementation. Given the significant financial component of the program, in addition to the more traditional environmental components, the occurrence of a multi-actor implementation arrangement makes intuitive sense. The variation in this variable also speaks to different paths states have taken in program implementation.

The variable for years of primacy also produced results generally as expected. Our findings here would suggest that states with a longer history of primacy direct more resources to these categories of applicants. The findings for NPDES authorizations provide evidence that the WQA has actually empowered states to implement the CWSRF program independently of congressional intent, thus suggesting that the goal to return programmatic authority to states has been achieved. States can not only choose whether (and when) to accept NPDES authorizations; the decision to accept this authority may not impact the ways in which states direct CWSRF resources to applicant communities. The decision to delegate permitting authority under the NPDES predates the WQA, yet several states have only accepted primacy in the past twenty years, and three states have continued to reject primacy. One of these states, Massachusetts, has significant water quality needs and has one of the larger CWSRF programs in the nation, which may serve to skew the findings to some degree.

Other variables provide additional insight into state allocation decisions. For example, leveraging dollars are especially important for environmental needs and small communities, but not for hardship communities or nonpoint projects. Given the funds required to meet these needs, leveraging enhances a state's ability to respond to significant environmental need. The same is true for small communities; given that EPA (2016) estimates some 80 percent of CWSRF resources are directed to small communities, this finding makes sense. By the same token, the added expense incurred in a loan program hampers the ability of the CWSRF to meet the needs of hardship communities; the negative relationship between leveraging and dollars for hardship communities illustrates this point. Likewise, the relatively low dollar value of the average nonpoint project means that leveraging does not provide any particular benefit for those needs. The effect of private sector actors in the CWSRF structure seems to have lessened over the years from the findings presented by Morris (1997), but the effect is generally consistent with our expectations across three of the four categories of CWSRF uses. Its effect is most obvious for dollars directed to small communities and hardship communities, where private actors have the effect of decreasing the dollars available for loans by almost $11 million per observation. Although the variable does not reach significance in the model for environmental need, the positive relationship is consistent with our initial expectations. If the private sector is the source of financial expertise, more complex financial systems with higher values are more likely to require advanced financial acumen to manage the funds effectively.

The results are a bit more mixed for the control variables in the model. Of the ten control variables, none were consistently significant across the set of models. On the other hand, median income failed to reach significance in any of the models. Centralization, professionalism, and political culture were significant in a majority of the models, but not always in the expected direction. The results were also quite mixed in terms of the party control variables. The best explanation for the findings for political control is that the funding decisions in the CWSRF, even though they are fundamentally political in nature, are made at an administrative level and are largely viewed by elected officials as administrative decisions rather than political decisions.

The measure of water quality needs also generally follows the expected pattern. Water quality needs are only calculated for point-source pollution, so one would not expect a significant relationship for nonpoint uses. In fact, as illustrated in the model for percent loans for nonpoint, the coefficient for water quality needs is negative and significant. For states with significant water quality needs, resources directed to nonpoint pollution are resources not directed to reduce point-source pollution. For the same reasons, perhaps more surprising is the lack of significance for water quality needs in the dollars for environmental need model. Since the water quality needs variable, by definition, measures the level of the most significant needs in the state, we might expect a positive and significant relationship.[20]

The negative coefficient for the NPS activism variable in our nonpoint models raises questions about the efficacy of the variable in broader applications. In spite of the encouragement of EPA to use CWSRF resources for nonpoint needs, the rates of use still fall short of CWSRF resources for other categories. As a broader measure of commitment to nonpoint pollution in the CWSRF over the life of the program, this variable leaves something to be desired in terms of its explanatory power. Perhaps a more robust measure of commitment to nonpoint-pollution control would yield better results. It is also possible that states do not yet see CWSRF resources as an important component of nonpoint-pollution policy.

Summary

Martha Derthick (1996, 3), a well-known scholar of federalism, addresses the future of federalism in the face of efforts in the 104th Congress to decentralize governmental authority. She writes:

Federalism both presumes and facilitates differences among the states. Assuming that the states are granted more freedom, will policy differences among them widen, or will their policies tend to converge?

In the case of the CWSRF program, a program specifically designed to grant states more policy freedom, state policy choices have clearly diverged, especially when viewed in comparison to the construction grants program, as specifically intended by the framers of the law. The wide variation in both program design and the distribution of program resources across the fifty states is evidence of this divergence. Still, the regulatory framework in the WQA, coupled with the "first use" requirement, serve as constraints in that states must enforce, or be subject to the enforcement of, water quality standards. The analyses in this chapter demonstrate the parameters of both the policy freedom and the constraints of the WQA.

Taken as a whole, the analyses presented in this chapter suggest a program for which the promises of Reagan federalism were, at least in part, met by states. The loan distribution patterns show significant variation across states, and suggest that states are making resource allocation decisions that are, at the least, consistent with state choice, even if that choice is contrary to the express desires of Congress. States are not hesitant to exercise discretion, not only in terms of the kinds of community they serve with the CWSRF program, but also how they design and administer the program. The same exercise in discretion can also be illustrated in the ways in which states fund, or fail to fund, clean water. A clear element of Reagan federalism was to move fiscal responsibility for some programs from the federal budget to the states. The expectation was that states would then assume not only program responsibility, but would allocate additional resources as well. These analyses indicate that while states have certainly assumed program responsibility, fiscal responsibility is another matter. Very few states have appropriated funds beyond the required state match for the capitalization grants, and still fewer than half of all states regularly leverage. The net result is that states are still highly dependent on continued federal funding for water quality. Given factors such as the state of the economy and constant pressure on state budgets, it seems unlikely that states will be positioned to assume full fiscal responsibility for clean water at any point in the foreseeable future.

The assumption of programmatic responsibility by states also has another important implication for public policy. Specifically, the delegation of program responsibility to states places the attainment of national policy goals in some jeopardy. While water quality issues are indeed being addressed by states, they are not necessarily being addressed in the manner envisioned by Congress in the WQA. In the case of assistance to hardship communities, it is likely that the nature of the policy instrument itself – a loan program – makes it especially difficult to meet these needs. On the other hand, our analyses appear to be consistent with EPA's (2016) conclusion that about 80 percent of CWSRF resources are directed to small communities – clearly consistent with the intent of Congress to serve these communities. Although it has taken some time to come to fruition, states are increasing the proportion of CWSRF resources directed to nonpoint needs. The lack of a valid and reliable measure of nonpoint-source pollution needs limits the ability to offer conclusions about program success, but the fact that states collectively continue to allocate resources for nonpoint projects is an encouraging trend.

Finally, program resources allocated for significant environmental need continue to account for the bulk of CWSRF funds. However, it is notable that, in spite of the investment made by states (and the national government) in this category, water quality needs continue to grow at a rate above total investment. This represents a troubling trend for the future, and one that will likely continue to grow. While general agreement exists that the state of the nation's waters is better now than it was in 1987, the increasing needs suggests the need for water quality funds will only grow. In an era of tight public budgets, and a host of unfunded (and under-funded) priorities in other policy arenas, it seems unlikely that the "needs" curve and the "resources" curve will intersect in the foreseeable future.

10

Promise and Performance

State Choice and National Water Quality Goals

The Water Quality Act (WQA) of 1987 stands today as the longest-lived expression of national water quality policy in the history of the nation. It is a product of a period of significant change in federalism, and represents a vision of federalism that is meant to reduce the role of the national government and increase state authority, responsibility, and discretion. The WQA's primary funding mechanism, the Clean Water State Revolving Fund (CWSRF) program, is designed to give states significantly more control over resources with which to meet water quality needs. Like other federal policies, though, the WQA is both liberating and constraining in terms of state action. While states undeniably have more discretion than was possible under the previous legislation, that discretion has clearly definable limits that govern state choice within the program.

The story of the WQA, and its predecessors, is one of largely incremental shifts in policy punctuated by more substantial shifts in approaches to clean water, all governed by the policy streams (Kingdon 1984) in place at the time. The first four decades of water quality policy followed a path of an increasing role for the national government at the expense of state discretion. From 1948 to 1972, the several amendments to the Federal Water Pollution Control Act imposed a greater level of federal involvement in all aspects of water quality, from funding to regulation. This span of time also saw a shift in the definition of water quality from a public health issue to an environmental issue. The 1972 legislation represents the zenith of federal control in water quality policy, a plateau that lasted roughly fifteen years. The WQA of 1987 symbolizes a significant shift in federal–state relationships in water quality that has remained in place for more than three decades.

The net result of this stream of policy activity is the initiation of more than 1,000 new loan agreements per year, on average, nationwide.[1] Allowable uses for CWSRF assistance have broadened over the years, and more states are funding projects that go beyond the construction of new wastewater treatment plants.

Although almost half the loans for centralized wastewater treatment made in FY 2016 were for Category I or Category II treatment, about 25 percent of loans that year were made for sewer system rehabilitation (EPA 2020a). Loans for projects such as energy and water conservation are still comparatively rare, but loans for nonpoint-source uses have increased dramatically in the last fifteen years. The results of these projects are difficult to quantify, but most observers agree the nation's waters are cleaner today than they were in 1987.

The focus of this research has been on state choice under the WQA, and the ways in which states addressed the requirement to design, implement, and administer their individual programs. In addition to the specific elements of state choice in the program, the contextual setting of each state was hypothesized to impact the degree to which in which states were able to meet the intent of Congress in terms of the distribution of program resources. States made choices based on the incentives, requirements, and limitations in the legislation, and those same incentives, requirements, and limitations are a direct result of the federalism goals of the framers of the legislation.

The Legacy of Reagan Federalism

The Water Quality Act represents something of a policy punctuation (see Jones and Baumgartner 2005) in the clean water policy arena. The inclusion of a grant-funded, state-administered revolving loan fund to replace the existing categorical grant program seems, in itself, relatively benign. However, in adopting this model, Congress fundamentally changed the role of the national, state, and local governments in water quality, and placed an unusually high burden on states in terms of program implementation. This was not simply a question of a new grant program, but the adoption of a new policy model for the provision of clean water. Although the framers of the law suggested that the federal role in funding should be time-limited to six years past the passage of the Act, the grant program has continued unabated a quarter century beyond the expiration of the original authorization.

In this sense, one might conclude that the WQA is something of a failure in terms of the tenets of Reagan federalism, which sought to end the federal fiscal responsibility for water quality. There is a grain of truth to this argument, in that federal funding continues. However, one might argue that other elements of Reagan federalism overwhelmed this goal. The main thrust of Reagan's vision of federalism was not only for reduced reliance on federal monies, but also for the devolution of program responsibility to states. In this regard, the WQA was wildly successful, in that the WQA empowered states to make programmatic choices on their own. Indeed, that very empowerment is largely responsible for the need to continue the federal grant funding. The framers of the WQA knew that the authorization in the Act would be insufficient to meet current or future water quality needs, and believed that states would either allocate additional state funding to the program, or leverage their programs to raise additional funds. States exercised their ability to make programmatic choices, and most states chose not to leverage. Even fewer states appropriated additional state funds to their state revolving fund. The reasons states took (or failed to take) these actions vary widely. But, this is the very essence of Reagan federalism: Let states make the choices that are best for the state, without undue federal pressure. Under the WQA, states did exactly that.

While other block grant programs were enacted during the Reagan years, and many of those are also still in operation, they differ from the CWSRF in that the CWSRF placed a level of administrative and financial responsibility on states unmatched by other block grants. States assumed unprecedented responsibility not only for the environmental aspects of the program, but also for a complex financial program that had to balance short-term needs with long-term financial viability. The financial balancing act was especially critical: If loan rates were too high municipalities would not be able to afford CWSRF assistance (or might find more competitive rates in the private bond market); if loan rates were too low, the fund would not remain financially viable in perpetuity. All of this responsibility happened under the broader mandate to meet water quality standards, or face sanctions and perhaps preemption by the national government.

The legacy of Reagan federalism, at least in terms of water quality, is thus reasonably unclear. While the CWSRF program indeed returned a great deal of programmatic authority to states, Reagan's concerns about costs, as expressed in both of his veto messages, have turned out to be prescient. The federal contribution to water quality has been greater in the thirty years since the passage of the WQA than in the thirty years preceding the Act. Likewise, another element of Reagan federalism, a desire to reduce federal regulation, was never intended by the framers of the WQA to be a feature of the legislation. If anything, the National Pollution Discharge Elimination System (NPDES) program was slightly strengthened in the WQA, and Title VI requires states to meet a variety of cross-cutting federal requirements for CWSRF assistance. And, although these appeared nearly a quarter century after passage of the Act, a Republican-controlled Congress added additional cross-cutting requirements to CWSRF funds[2] that went beyond the several cross-cutting requirements in the original legislation.

Reagan's concerns about the nonpoint-source pollution provisions of the WQA were also somewhat prescient, albeit for reasons related directly to his desire for programmatic devolution to states. Reagan feared new federal regulation that would govern nonpoint pollution. Instead, a lack of federal action to emphasize the nonpoint sections of law allowed many states effectively to ignore nonpoint pollution. Ultimately it was pressure from the courts, itself a result of suits brought by citizen groups against both the Environmental Protection Agency (EPA) and states (see Houck 2002) that led states to develop nonpoint-pollution control strategies. The distribution patterns of CWSRF resources reflect this pattern. Prior to the mid-2000s few states funded nonpoint projects in the CWSRF, but in the period since the use of CWSRF resources for nonpoint projects has increased substantially.

Has the Water Quality Act and the CWSRF Met Expectations?

The fundamental question behind this research is whether state actions under the Water Quality Act have met the intent of Congress. The "intent of Congress" is, in many ways, hard to determine, since Congress itself is a body of diverse interests and preferences. However, it is not unreasonable to define "intent" as the set of expectations found in the WQA itself. To this end, we treat the expectations of Congress to include the four uses of CWSRF funds discussed in the Act: to meet significant environmental need; to meet the needs of small communities; to meet the needs of financially at-risk (hardship) communities; and to address nonpoint projects.

For the category of environmental need, the data allow us to conclude that the CWSRF program has had its greatest success to date in this category. The bulk of CWSRF resources have been directed at environmental need, and thousands of projects around the nation have been built using CWSRF funding. Although needs continue to grow, the gap between needs and resources has remained largely steady over the life of the program. The WQA has not "solved" the problem of environmental need, but it has largely kept pace with the growing need. Environmental need is not a stationary target; it grows as the population increases and as infrastructure built in previous decades reaches the end of its useful service life. As the bulk of CWSRF resources have been dedicated to applicants with significant environmental need, we may reasonably conclude that the CWSRF program has been largely successful in meeting that goal.

The same may be said of the desire of Congress for the program to serve small communities. While challenges remain regarding the affordability of loan assistance for systems with a relatively small number of rate payers, it is also clear that states have successfully directed CWSRF funds to small communities. This is an element of the program that has taken some time to come to fruition; Heilman and Johnson (1991) and Morris (1997) reported that few states were directing resources to small communities in the early years of the program.

As EPA stated in its 2018 annual CWSRF report, by 2017 some 80 percent of CWSRF resources had been directed toward small communities (EPA 2018).

A similar pattern may be seen in the resources directed to nonpoint pollution. By 1994, less than one-sixth of all loans had been made for nonpoint pollution, but by 2016 nearly 46 percent of all loans had been made for nonpoint projects. The value of these loans tends to be relatively small compared to centralized wastewater treatment projects, but only because nonpoint-pollution projects tend to be less expensive to build than, for example, treatment plants to provide Category I or Category II treatment. We can conclude from this that CWSRF resources have been successfully employed by states to address nonpoint-pollution needs in states.

A very different picture emerges from efforts to address the needs of hardship communities. Between 1988 and 2016, less than 10 percent of all CWSRF loans had been made to hardship communities, and fewer than 7 percent of total loan dollars distributed went to these communities. While this percentage is not encouraging, the reason likely does not lie in the pattern of state decisions, but rather the underlying logic of the CWSRF model. A loan is very different than a grant, in that a loan (even a loan at zero percent interest) must be repaid. For a community already strapped for cash, or with a significantly lower median household income, the resources to pay loan principal and interest are limited. States are likewise caught in a bind, in that loans made below the prevailing interest rates threaten the perpetuity of the fund itself. Some states have even resorted to the use of negative interest rates[3] (see Heilman and Johnson 1991; Bunch 2008); while this practice no doubt provides a welcome subsidy to the applicant community, it also has deleterious effects on the long-term value of the fund. Although little evidence of this exists at the national level, it seems likely that Congress' actions to appropriate Title II funds well beyond the expiration of the authorization was an attempt to address this need. In sum, the unmet needs of hardship communities are likely a function of the loan model itself.

The lack of a useful measure of program outcomes is a significant limitation to any analysis of the CWSRF program and, by extension, of the Water Quality Act. In the opening chapter of this book, we suggested that environmental programs could take one of two possible approaches to environmental quality (or a combination of the two). One approach is to focus on the application of technological controls to reduce the pollution stream. In many respects, this has been the focus of not only the CWSRF program, but of all federal water quality legislation since 1948. By setting effluent standards and providing states with funds to help build the infrastructure, the focus is squarely on pollution control technologies. The second approach, setting ambient water quality standards, was also included in the WQA, albeit as something of an afterthought. Indeed, it took a series of lawsuits by citizens and environmental groups to force the EPA to enforce the total maximum daily load (TMDL) provisions of the WQA.[4] Testing is expensive, especially in states with many miles of rivers and streams, and/or many lakes. The lack of testing resources, coupled with a lack of incentive to develop a more comprehensive testing program, limits our ability to tie program activity to program outcomes. Without comparable measures of ambient water quality across states, we are largely limited to an examination of short-term program outputs.

As an example of this issue, a survey of state CWSRF coordinators conducted in 2020 in preparation for this research asked respondents how they measured program outcomes. The most common response (twelve states) was the number of loan agreements entered into each fiscal year; the second-most common response (four states) was by the number of pounds of nitrogen removed at each wastewater facility. At least two respondents remarked that the question did not apply in the context of the CWSRF program, because the goal of the CWSRF program was to make loans, not enhance water quality. While the latter responses might be dismissed as nearsighted, they demonstrate the underlying dilemma of federal water quality policy: Everyone agrees water quality is both important and desirable, but measuring changes in the quality of the water is challenging. In the context of the CWSRF program, with its strong focus on point-source pollution, measuring changes in ambient water quality are further complicated by the growing realization that nonpoint-pollution abatement must be a major element of any comprehensive water quality goal. The TMDL requirements are a realization of this issue, but the lack of resources to reduce nonpoint pollution (in the CWSRF program and elsewhere), coupled with the lack of a common water quality testing regime, makes this perhaps the greatest weakness of federal water pollution policy.[5]

Water Quality Policy and the Changing Face of Federalism

As this chapter is being written, following the presidential election of 2020, questions swirl regarding the future of water quality policy. On the one hand, demand for clean water continues largely unabated, and significant unmet needs for water quality funding remain. It seems likely that Congress will continue to appropriate funds for Title VI. Infrastructure projects are generally popular in Congress, and previous Congresses have shown no hesitation to continue to fund the CWSRF absent an active congressional authorization. The program is also no doubt popular in states, and provides jobs in addition to clean water.[6] Every dollar sent to a state from the national government represents, in effect, one dollar less the state needs to allocate to water quality. Although a few states have declined a capitalization grant over the years, the flexibility afforded to them in how they raise the matching funds has made it easier for states to meet the match requirements and thus made the decision to accept the capitalization grant a relatively easy one.

Against this backdrop, however, is a chaotic combination of an unsettled political environment, divided government, and rampant hyperpartisanship. The Senate has been especially passive over the past decade, and the majority leader for much of that time, Senator Mitch McConnell (R-KY), has refused even to consider legislation across a wide spectrum of policy arenas. Given this environment, it is doubtful that a revision to the Water Quality Act (and thus a new authorization) is probable in the foreseeable future. A Congress preoccupied with a worldwide pandemic, a stagnant economy and high unemployment, and myriad other demands, is unlikely to place water quality high on the agenda, even absent the hyperpartisanship. While plans for infrastructure policy have been circulated in Congress for years (see Morris, Williamson, and Meiburg 2020), they have just come to fruition. As this book goes to press, Congress has just passed a large infrastructure bill that authorizes in excess of $1 trillion for infrastructure needs, including $55 billion for water needs. Notably, the legislation was approved with bipartisan support. In its final form, the law authorizes about $2.5 billion per year over five years for water quality needs, or roughly a fourfold increase above current funding levels. The legislation also reduces the state match to 10 percent and earmarks additional funds for loan forgiveness and grants, a potential boon for small and hardship communities.

The last four years have also witnessed something of an assault on federalism (see Bowling, Fisk, and Morris 2020, among others), as former President Trump effectively "weaponized" federal assistance to favor some states at the expense of others. It appears that the resources involved were largely limited to emergency assistance, but the uncertainty surrounding the arbitrary distribution of federal assistance calls into question the underpinnings of American federalism. Other scholars have described the recent practice of federalism as "coercive" (Kincaid 1990), "partisan" (Bulman-Pozen 2014), "fragmented" (Bowling and Pickerill 2013), or "transactional" (Bowling, Fisk, and Morris 2020); these terms all speak to a period of conflict and, likely, stagnation in federal–state relationships.

Environmental policy has always been something of a flash point in American politics, and preferences about the line between benefits and costs tends to be polarizing in the best of circumstances. Although water quality policy has generally enjoyed a high level of support in Congress, the policy arena has not been without conflict. Liberals tend to favor greater environmental regulation, while conservatives tend to favor less regulation. For example, the Obama administration worked for several years to develop a new "Waters of the US" (WOTUS) rule to define the meaning of "navigable waters" (and thus what waters were subject to federal regulation). However, the WOTUS rule was rescinded by the Trump administration in 2019, a move that followed the fate of scores of other environmental regulations during the Trump presidency. As the nation grows more partisan, these whiplash reversals in policy are more likely, and the prospects for new or updated legislation become concomitantly less likely.

Implications for Research and Policy

The research in this volume has focused specifically on the policy theory of the CWSRF program, and the assumptions of Reagan federalism in terms of the ability of the CWSRF to serve specific categories of applicants. Taken in total, there are four major implications for both research and public policy in the water quality arena. First, the assumption that states have the capacity to assume responsibility for complex federal programs is clearly not met in all cases. Without a doubt, some states were ready, willing, and able to assume responsibility for the CWSRF program. Other states were less ready; thirty years later, some states still lack the resources necessary to implement the program fully. It would also be incorrect to assume that all states desired to adopt the CWSRF model; responses from the early survey by Heilman and Johnson (1991) indicated that nearly 10 percent of states preferred the Title II program (construction grants) to Title VI, and nearly a quarter of respondents expressed doubt about the longevity of the CWSRF program. More importantly, our analyses suggest that states with insufficient resources at the time of initial implementation lagged behind other states in terms of the distribution of loans to small communities, hardship communities, and for environmental need, although those differences have become less important over the years since initial implementation. Still, the initial conditions in states largely determined both program structure and program implementation, and the effects of those decisions are still present thirty years later. In short, the false assumption of initially sufficient resources across all states has effects that continue to be detected today. For scholars of implementation (and of federalism), this suggests models of state implementation should include measures of initial capacity. The effects of a lack of capacity may well be less (or more) important in other policy arenas, but that is a statement best answered empirically.

The policy implications of this issue bring to the fore policy choices driven by different conceptions of federalism. If a fundamental assumption is not supportable, it brings into question the ability of states to achieve the policy goals with which they have been charged. While states can (and do) actively resist federal policy mandates, state objections are more typically ideological in nature. Federal mandates to be addressed by states, particularly in complex policy arenas, might prove more successful if federal agencies were able to provide adequate resources, including training. This issue is not meant to ignore the balance between national policy and state sovereignty, but rather to suggest that national mandates administered by states should take into account inevitable differences in state capacity.

Second, there is a clear trade-off between national policy goals and state discretion. The CWSRF provides a great deal of flexibility on the part of states, both in terms of program structure and the distribution of resources from that program. When the national government cedes control of the distribution of resources to states, they necessarily place national goals in a subservient position to state policy preferences. There are certainly reasonable arguments for such a pattern; a common argument is that states know what is best for states than do policy makers in Washington, DC. While there is clear merit to that argument, there is also an equally reasonable counterargument that national goals are best achieved when all states put forth maximum effort. This may be especially true for environmental problems, which do not respect political boundaries.[7] Still, policy makers should carefully consider the trade-offs between the desire to attain national goals and a desire to delegate responsibility for those goals to states, particularly in the case of complex policy instruments.

Third, and relatedly, the balance between national policy goals and state incentives will tend to favor state interests when states are given primary responsibility for program implementation. At one level, this is a natural outcome of state discretion and states' rights. The United States is both geographically large and diverse, and the differences inherent in "place" matter. As Elazar (1966) observed more than fifty years ago, state political culture matters, and policy preferences differ from state to state. Woodward (2011) makes largely the same argument, suggesting that regional cultures are even more powerful explanators of state behavior. The debate over states' rights and federal power has been an enduring feature of American governance since the First Continental Congress (Kettl 2020), and it continues to the present day. Although the balance between the two has ebbed and flowed over the years, state governments have been remarkably consistent in placing state interests ahead of national interests.

Finally, revolving loan funds have the potential to provide stable, long-term resources for a range of uses, and they are especially attractive as policy instruments to meet capital-intensive infrastructure needs. It is not a surprise, for example, that Congress has also employed the revolving loan model to meet drinking water infrastructure needs, and the model has garnered support from many groups as a solution for transportation infrastructure as well. However, it is equally clear from the water quality experience that revolving funds can only meet their long-term potential if they are adequately capitalized. Congress and EPA realized that the original authorization for the CWSRF would fall well short of the amount necessary to provide sufficient resources to meet state needs. The solution to this problem was to encourage states to leverage their capitalization grants, and to encourage states to appropriate additional funds to the CWSRF. Few states took the latter approach, and those that did allocate additional funds did so sporadically. However, even states that have leveraged aggressively, and have done so with capitalization grants well past the initial authorization, have failed to meet all of their water quality needs. If federal capitalization had ended as scheduled in 1994, and if every state had leveraged aggressively, most states would still have fallen well short of their needs. Moreover, the same political forces that sought a reduction in federal spending also sought reductions in state spending. Most states are also prohibited by their constitutions from engaging in deficit spending, and billions of dollars in additional funds for water quality are unlikely to be provided by cash-strapped legislatures. While there is little doubt that the CWSRF has provided tens of billions of dollars for clean water, without federal funding beyond the initial authorization it is likely that program outcomes would be much less impressive.

It is also worth noting that any analysis of the CWSRF program, and of the WQA, is necessarily incomplete without a detailed analysis of the behavior of municipal governments. Local governments are the intended target for CWSRF loans, and local governments operate the overwhelming majority of wastewater treatment plants. The CWSRF program provides a specific set of incentives (and disincentives) for participation, and the decision process at the local level that leads some communities to decline to participate in the CWSRF program is important to our understanding of the program itself. Unfortunately, collecting meaningful data for analysis from local governments nationwide is a significant undertaking. It is instructive that there are no published empirical studies of local governments and the CWSRF program in the extant literature, and federal government agencies (e.g., EPA, the US Government Accountability Office, and the Congressional Research Service) have been largely silent on the issue.[8] The most useful study would be one that included both communities that have received CWSRF assistance, along with a sample of communities that have not received assistance. To collect nationwide data for such a study would require a significant investment in both time and money, and would require a dedicated effort on the part of a large team of researchers. However, the local government perspective is ultimately critical to a more complete understanding of the CWSRF program. It is highly likely that local government behavior helps determine state behavior, and vice versa.[9]

Summary

This book begins with an epigraph by French author, Victor Hugo, ruminating on the role of the sewer in Parisian society. If Hugo is correct that the sewer is, indeed, the conscience of the city, then we might extrapolate from that statement that clean water is likewise the conscience of a nation. Citizens both need and demand clean water, and governments have responded to the demand with a series of policies designed to provide clean water. However, like so many other policies, "the devil is in the details." The question of who pays for clean water is especially difficult given the disconnect between the high, concentrated cost of clean water and the distributed benefits of clean water. How society chooses to address clean water, and how it apportions both costs and benefits for clean water, speaks to the very nature of society. In its retrospective of the Clean Water Act, AISWPCA (2004, 27) writes the following:

Though much work remains to be done, it is clear that the Clean Water Act has paved the way for remarkable accomplishment. Clean water is essential for good health, and water resources are critical to much of the economy, including the recreation and tourism industry, agriculture, and the fish and shellfish industry. The Clean Water Act has been an investment in the nation's economy. It has been an investment in the health of American citizens. And it has been an investment in America's future.

There is little disagreement that the Clean Water Act has resulted in a significant investment in water quality infrastructure, and that the result has been an improvement in the quality of the nation's waters. In the years since AISWPCA made their observations, many more billions of dollars have been invested in water quality infrastructure; yet it is still the case that much work remains to be done.

The Water Quality Act has been the law of the land for more than thirty years, and it represents a fundamental shift in federal–state–local relationships in water quality. From the early days of water quality as a traditional pork barrel program, water quality has grown to become a serious policy effort on the part of the national government. The policy instruments (Bressers 1992) in place at any point in time reflect the current thinking about the proper role of the national government in the provision, production, and delivery of a good or service. As the problem, solution, and politics streams (Kingdon 1984) change, so do the policy alternatives both available and possible at any point in time. If clean water is indeed essential for human health, the economy, and recreation (AISWPCA 2004), then its place as the conscience of the nation is paramount.

Appendix A

List of Variable Descriptions and Data Sources

Dependent Variables

The dependent variables consist of data on CWSRF loans taken from the National Information Management System (NIMS) maintained by EPA as the repository of CWSRF national- and state-level data. The NIMS covers data from FY 1988 to FY 2019 (as of this writing in late 2020), and consists of over 300 discrete variables. The data are available in Portable Document Format (pdf) files, by state, at www.epa.gov/cwsrf/clean-water-state-revolving-fund-cwsrf-national-information-management-system-reports.

The data for the dependent variables employed in this study are the dollar values for loans for Category I and Category II needs (significant environmental need); communities with populations less than 10,000 (small communities); total dollar value of nonpoint projects; and dollar value of loans to hardship communities (financially at-risk communities) for each fiscal year from 1988 to 2016. All values are adjusted for annual inflation; the inflation factors are provided by the US Bureau of Labor Statistics (2020).

Loan percentages for each category of uses is calculated by summing the total of all loans made in each state per fiscal year, and then dividing that figure by the number of loans made for each of the dependent categories.

It is important to note that states may report negative dollar values in a given year. These numbers are the result of account adjustments that occur as the result of fund adjustments, audits, and fund reapportionments.

Independent Variables

The independent variables in the study are derived from several sources. Table A.1 lists the variable, the definition, and the data source.

Table A.1 *Independent variable definition and data source*

Variable	Description	Data Source
Program location	The primary contact agency for the CWSRF program by type (environmental/natural resource, health, financial).	NIMS; lists of State CWSRF contacts archived by EPA at epa. gov; Heilman and Johnson (1991)
Leveraging dollars	The net dollars raised through leveraging (line 209), adjusted for inflation.	NIMS (EPA 2020a)
Number of state agencies in CWSRF implementation	The number of state agencies reported with a direct role in CWSRF operations.	Heilman and Johnson (1991)
Initial state sufficiency of resources	Survey responses from questions that asked about initial sufficiency of environmental and financial personnel expertise.	Heilman and Johnson (1991)
Year of state water quality primacy	The year in which a state was first granted primacy by EPA in the area of water quality.	EPA
NPDES score	The number of areas in which a state has assumed primary responsibility under the NPDES.	Hoornbeek (2011)
Privatization scores	Factor scores derived from survey data regarding the extent of private sector involvement in CWSRF implementation. Scores for formal and informal privatization are summed to create a single privatization score for each state	Morris (1994)

Control Variables

The control variables for this study are drawn from a wide range of sources. Table A.2 details the variable name, description, and source.

Table A.2 *Control variable definition and data source*

Variable	Description	Data Source
Demographic		
State population density	The number of people in a state (state population), divided by the number of square miles of state territory.	American Community Survey (ACS; US Census 2020)
Median income	The median household income in a state, adjusted for inflation.	American Community Survey (ACS; US Census 2020)
Political		
Government ideology	A measure of the ideology of state elected officials, including both state and national offices.	Berry et al. (2010)
Legislative professionalism	A measure of the relative professionalism in a state legislature that includes salary and benefits, time demands of service, and staff and resources.	Squire (2007)
Political culture	A measure of a state's general approach to public policy. The three categories are moralistic, individualistic, and traditionalistic. In these analyses traditionalistic is used as the reference category.	Elazar (1966)
State centralization	A measure of the relative dominance of state government compared to local governments in the state.	Bowman and Kearney (2011)
Democratic/ Republican Trifecta	Whether a single party controls all branches of state government (House, Senate, and governor's mansion).	NCSL (2020)
Environmental		
Water quality needs	A quadrennial estimate, in dollars, of the value of water quality infrastructure needs in the state. The data are compiled by states and sent to EPA for inclusion in a report to Congress. The most recent report available is 2012. The data for the three years following the date of each report are assumed to be the same as for the year in which the report was issued. All values are adjusted for inflation.	EPA (1989b, 1993, 1997b, 2003, 2008, 2010, 2016)
NPS activism score	A scale designed by Hoornbeek (2011) that measures four elements of state legislation to address nonpoint pollution. The variable can range from 10 (complete regulation across all four areas) to 0 (no significant enforceable regulations).	Hoornbeek (2011)

Appendix B

Descriptive Statistics and Correlation Table for Variables in Regression Models

Table B.1 *Descriptive statistics for independent, control, and dependent variables*

Variable	Obs.	Mean	Std. Dev.	Min	Max
Independent					
Institutional location	1,352	1.568787	1.13409	1	6
Leveraging dollars*	1,450	37.4013	108.1648	−3.791111	1,317.244
Total actors	1,450	3.62	1.777014	1	8
Initial sufficiency – Env.	1,450	0.74	0.4387856	0	1
Initial sufficiency – Fin.	1,450	0.84	0.3667325	0	1
Years of primacy	1,450	36.74	14.01313	0	47
NPDES authorizations	1,450	3.6	1.183624	0	5
Privatization factor	1,276	0.0385876	1.16142	−1.23738	4.2927
Control					
Demographic					
Population density	1,450	184.4713	250.5864	0.9492101	1,218.047
Median income	1,450	63,701.97	10,140.82	39,240	91,712
Political:					
Government ideology	1,450	50.26431	15.23374	8.44989	97.0015
Centralization	1,450	47.04083	7.8515	35	78
Traditional	1,450	0.32	0.4666371	0	1
Moralistic	1,450	0.34	0.4738722	0	1
Individualistic	1,450	0.34	0.4738722	0	1
GOP Trifecta	1,450	0.2441379	0.4297231	0	1
Dem Trifecta	1,450	0.2365517	0.4251113	0	1
Professionism	1,450	0.1963938	0.1252128	0.027	0.659
Environmental					
WQ needs*	1,450	1,651.677	2,763.352	0	19,042.87
NPS activism score	1,450	12.45	7.762349	2	25.5
Dependent					
Environmental need $*	1,450	56.29737	95.95987	−3.835547	1,151.837
Environmental need %	1,450	37.17568	25.59968	−20	100
Small community $*	1,450	22.58918	29.26229	−3.689901	333
Small community %	1,450	43.33473	28.39051	0	100
Hardship community $*	1,450	8.349549	25.59471	−8.195717	239.7395
Hardship community %	1,450	7.74849	16.84339	0	150**
Nonpoint project $*	1,450	3.712115	15.26868	−16.80246	322.4575
Nonpoint project %	1,450	12.52159	24.41257	0	100

*In millions of dollars.
**Number is greater than 100 due to double-counting.

238

Table B.2 *Correlation matrix of all continuous independent and control variables*

	State Leverage Match	Total Actors	Years of Primacy	NPDES Authorizations	Privatization Factor	Population Density	Median Income	Government Ideology	Centralization	Professionalism	WQ Needs	NPS Activism Score
State leverage match	1											
Total actors	−0.1464	1										
Years of primacy	0.1122	0.1263	1									
NPDES authorizations	0.026	0.1058	0.3433	1								
Privatization factor	0.0249	0.3826	0.0338	0.2486	1							
Population density	0.185	0.0743	0.2158	0.0071	0.2094	1						
Median income	−0.0079	0.0184	−0.0061	−0.0108	0.0085	−0.0033	1					
Government ideology	0.1524	−0.1031	0.2495	−0.046	−0.0801	0.5057	−0.0197	1				
Centralization	−0.2015	−0.139	−0.1004	−0.1038	−0.035	0.2022	0.0096	0.3391	1			
Professionalism	0.3461	−0.0004	0.2497	0.1992	0.0165	0.2146	0.0234	0.2703	−0.292	1		
WQ needs	0.465	0.0301	0.1193	0.0651	−0.0078	0.2292	−0.0056	0.1593	−0.4109	0.649	1	
NPS activism score	0.1364	0.031	0.2495	−0.0007	−0.2232	0.4796	−0.0153	0.4918	−0.1566	0.3929	0.3514	1

n = 1,276

About the Author

Dr. John Charles Morris is a professor in the Department of Political Science at Auburn University. Prior to his appointment at Auburn, Dr. Morris held a joint post as a professor of public administration in the School of Public Service at Old Dominion University, and professor of political science in the Department of Political Science and Geography. He also served as an associate professor in the Department of Political Science at Mississippi State University. He has studied environmental policy and water policy for more than twenty-five years, and has published widely in public administration and public policy. He is the co-editor of five volumes, including *Speaking Green with a Southern Accent: Environmental Management and Innovation in the South* (Lexington Press, 2010), *True Green: Executive Effectiveness in the US Environmental Protection Agency* (Lexington Press, 2012); and *Advancing Collaboration Theory: Models, Typologies, and Evidence* (Routledge, 2016). His most recent books include *The Case for Grassroots Collaboration: Social Capital and Ecosystem Restoration at the Local Level* (Lexington Press, with William Gibson, William Leavitt, and Shana Jones, 2013); *State Politics and the Affordable Care Act: Choices and Decisions*, co-authored with Martin Mayer, Robert Kenter, and Luisa Lucero (Routledge, 2019); *Organizational Motivation for Collaboration: Theory and Evidence*, co-authored with Luisa Diaz-Kope (Lexington Press, 2019); and *Multiorganizational Arrangements for Watershed Protection: Working Better Together*, co-authored with Madeleine Wright McNamara (Routledge, 2020). His most recent book is *Policy Making and Southern Distinctiveness*, co-authored with Martin Mayer, Robert Kenter, and R. Bruce Anderson (Routledge, 2022). In addition, he has published more than seventy articles in refereed journals, and nearly forty book chapters, reports, and other publications. His work appears in journals such as the *Journal of Politics, American Review of Public Administration, Policy Studies Journal, Voluntas, Publius: The Journal of Federalism*, and *Public Administration Review*, among others.

Dr. A. Stanley Meiburg became the Director of Graduate Studies in Sustainability at Wake Forest University in Winston-Salem, North Carolina, in 2017, following a thirty-nine-year career with the US Environmental Protection Agency. From 2014 to 2017, Dr. Meiburg served as EPA's Acting Deputy Administrator, the agency's second highest position. Before becoming deputy administrator, Dr. Meiburg served in senior career positions as EPA's Deputy Regional Administrator in the Southeast and South Central regions of the United States, as well as in EPA offices in Research Triangle Park and in Washington, DC. He received EPA's Distinguished Career Service Award, EPA's Gold Medal for his work on the Clean Air Act Amendments, the Commander's Award for Public Service from the Department of the Army, and the Distinguished Federal Executive award, the highest civilian award for a Federal senior executive. He is currently chair of the North Carolina Environmental Management Commission, a member of the Roundtable on Science and Technology for Sustainability of the National Academies of Sciences, Engineering and Medicine, and a Fellow of the National Academy of Public Administration. Dr. Meiburg holds a B.A. degree from Wake Forest University, and masters and doctoral degrees in political science from Johns Hopkins University.

Notes

Chapter 1

1 Rachel Carson's *Silent Spring* (1962) is widely considered to be one of the most influential statements of the public nature of environmental problems. It also served as a rallying point for many people in the early years of the environmental movement in the United States.

2 The short title "Clean Water Act" did not appear in law until the 1977 amendments to the FWPCA. However, the term "Clean Water Act" is often applied to the 1972 legislation (see Adler, Landman, and Cameron 1993; see also Davis 1998, 6). For example, EPA Administrator Carol Browner delivered a speech in Minneapolis, MN, on October 17, 1997, titled "25th Anniversary of the Clean Water Act, Minneapolis, Minnesota" (EPA 1997a). To avoid confusion, we will follow that convention in this book.

3 The term "policy instrument" refers to a tool, process, practice, method, or solution used to achieve a desired policy outcome. For a more detailed discussion, see Bressers and Klok (1988), and Bressers (1992).

4 See Heilman et al. (1994) for a fuller discussion of the history of water quality policy in the United States.

5 I am indebted to Dr. Gerald A. Emison for helping me understand the link between municipal growth management and water quality policy. Although the issue is addressed only briefly in this volume, it is an issue worth of further consideration and scholarly exploration, particularly in terms of its effects on local governments.

6 A third potential choice might be to ignore effluent standards and discharge untreated effluent into the waterways. However, the regulatory framework in place (the National Pollution Discharge Elimination System, or NPDES) provides for substantial ongoing financial penalties and fines and, in extreme cases, prison sentences for polluters. These fines can be as high as tens of thousands of dollars per day, which provides a strong incentive to avoid illegal discharges.

7 Other classes of applicants, such as Native American tribes, were also included in the list. However, the relative needs of these groups were smaller, and as a practical issue, little in the way of CWSRF funds have been directed to these applicants. This issue will be further explored in later chapters of this book.

8 This issue has proven to be one of the more contentious elements of the WQA. A number of states have made CWSRF loans for nonpoint pollution, but the overall application has been uneven. Moreover, beginning in the new century, several environmental groups began to sue the EPA to force compliance with regulations for nonpoint-source pollution. This led to the development and implementation of Total Maximum Daily Load (TMDL) plans to address nonpoint-source pollution in waterways. This in turn led to a number of states and other groups suing the EPA to prevent enforcement of the TMDL standards. See Houck (2002) for additional detail; we will also address this issue more fully later in this volume.

9 The question of the specific category of grant represented by the CWSRF program is the subject of some debate. As I argue later in this chapter, the CWSRF grant shares most of its characteristics with those of block grants.

10 Bowling, Fisk, and Morris (2020) note that the chaotic response of the Trump administration to the pandemic is indicative of a failure of leadership, exacerbated by a lack of vision about appropriate intergovernmental relationships. Employing the term "transactional federalism," Bowling, Fisk, and Morris (2020) provide several examples that illustrate the tensions of federalism in a hyperpartisan political environment.

11 There are two issues at play here. First, while some nonpoint pollution can be traced to a source, in many instances the specific polluter is unknown (or the costs of tracing the pollution would be too great). Second, within the limits of technology, it would be possible to create a permitting process for nonpoint pollution. However, the choice to exempt nonpoint-source pollution from the permitting process is also a policy choice, and to date US clean water policy has chosen to exempt nonpoint pollution from the permitting process.

12 Indeed, Congress spent much of the next ten years rolling back the "no-discharge" requirements in the CWA. They were abandoned completely in the 1987 legislation. This issue is examined in greater detail in Chapter 4.

13 The WQA included a program that required states to develop TMDL standards for waterways within the states. The intent was to focus on ambient standards, and encourage states to address both point-source and nonpoint-source pollution in the waterways. This program was largely dormant until some states and citizen groups sued the EPA to force compliance with these requirements (see Houck 2002), and the outcomes of this pressure are not yet clear. The TMDL requirements have been strongly opposed by farmers, among others.

14 Although funding for construction grants was authorized in the 1948 legislation, no funds for this purpose were ever appropriated for that legislation. However, with the 1956 amendments to the FWPCA, Congress both authorized new spending and appropriated funds for construction grants. The authorizations increased in every iteration of the FWPCA from then on. This issue is discussed in greater detail in Chapters 3 and 4.

15 The WQA also contained an authorization for Title II funds (construction grants). Authorized for a period of two years, the intent was to provide states with a "bridge" program to maintain a funding stream as states prepared for the new CWSRF program. Although authorization expired in 1990, Congress continued to appropriate Title II funds through FY 2013.

16 See National Council on Public Works Improvement (1988); see also Heilman and Johnson (1992, chapter 2) for an analysis of WTW funding patterns during this period.

17 The funding options considered by the task force included (1) federal loans, (2) federal grants, (3) municipal financing, (4) municipal bonds with credit enhancements offered by the federal government, and (5) capitalization grants from the federal government for state revolving funds (EPA 1984).

18 The WQA allowed for a transition between the construction grants program and the new revolving loan fund program. The legislation included authorization for additional construction grants to be phased out between FY 1987 and FY 1990, at which time authorization for the construction grants program would expire.

19 Previous research (Heilman and Johnson 1991) indicates that most states used a standard twenty-year amortization period for CWSRF loan assistance. Later legislative changes gave states the option to extend the amortization period to a maximum of thirty years.

20 Other popular institutional locations for the CWSRF program include state departments of health or state-run financial agencies.

21 As will be developed later in this book, the WQA does allow states to use CWSRF resources to address nonpoint pollution, as well as other uses (water reclamation, desalination, etc.). However, it is also the case that the overwhelming portion of CWSRF resources have been directed at point-source projects, in the form of the construction wastewater treatment plants and their associated collection systems.

22 This shift toward state discretion (or state primacy) was tested by Crotty (1987), who noted wide variation in states' willingness to undertake responsibility for the implementation of federal environmental standards.

23 See Heilman and Johnson (1991) for a further discussion of the scope of the changes affecting the role of water quality personnel in the states as a result of the shift to the CWSRF model.

24 It is important to note that, even with the creation of the NPDES permitting process, states still retained a significant role in water quality. States could accept primacy, under which states were given primary responsibility to determine water quality standards in the state, and became the primary regulator of water quality. However, states that did not accept primacy, or who failed to meet federal requirements, would effectively cede that authority back to the national government. States were also free to set water quality standards to be more stringent than the national standard. These issues will be addressed in greater detail in later chapters in this book.

25 The notion of "policy streams" was developed by Kingdon (1984) as a means to describe the development and movement of policy issues, alternatives, and decisions in the broader social, political, and economic contexts of the policy arena.

26 The issues surrounding leveraging are discussed in greater detail later in this volume.

27 This figure includes an additional $3.9 billion appropriated in FY 2009 under the auspices of the American Recovery and Reinvestment Act (ARRA) enacted as an economic stimulus following the market crash of 2007–2008. The total CWSRF appropriation for that year was $4.3 billion.

28 The EPA's National Information Management System reports a total of roughly $46.58 billion in federal capitalization grants, and $8.53 billion in required state matching funds. However, with leveraging and additional state funding, the total CWSRF investment to date has been on the order of $138.45 billion (EPA 2020a).

29 The term "medium" is not defined in the report, but rather is meant to indicate a midpoint along a continuum of grant types. The other two categories listed are "low" and "high," suggesting that "medium" falls somewhere between these two endpoints.

30 Metapolicy is defined here as a *policy about policymaking*. See Dror (1968).

31 As detailed later in this volume, the transition from categorical grants to block grants was not abrupt. The original intent of Congress was to provide a transition period during which states would be eligible for both categorical grants and block grants. Title II funding was authorized through FY 1990, at which time authorization would expire. Although authorization did indeed expire, it is important to note that Congress continued to appropriate funds for Title II grants through FY 2013.

Chapter 2

1 OMB Circular A-76 was first issued in 1966, and revised in 1967, 1979, 1983, 1996, and 1999. In short, A-76 suggests that government should not compete against the private sector and, wherever possible, the private sector should be used as a source of goods and services by government. The policy calls for agencies to perform a cost-benefit analysis to determine whether money can be saved through contracting. This policy became the basis of Reagan's expanded efforts to privatize many of the more traditional functions of government. See Kettl (1988).

2 In 1980, Reagan garnered 489 electoral votes across forty-four states to Carter's forty-nine votes across six states. Reagan also won 50.7 percent of the votes against Carter's 41.0 percent (third-party candidate John Anderson received roughly 6.6 percent (Peters and Woolley 2020). In the 1984 election Reagan won all but eight electoral votes (Mondale won only his home state of Minnesota). Reagan also secured 58.8 percent of the popular vote, compared to Mondale's 40.6 percent (Peters and Woolley 2020).

3 Kingdon's (1984) term "policy stream" is applied here as a heuristic to direct attention to a set of seemingly unrelated events that acted to shape the water quality policy arena in the 1980s.

4 An important element of this shift was a concomitant shift from the use of categorical grants toward block grants. As detailed later in this chapter, block grants gave states greater freedom in terms of how to apply the grant money within their state.

5 Watson and Vocino (1990, 430–431) note the fact that the Tax Reform Act had the effect of reversing the trend toward privatization, in spite of the Reagan administration's enthusiastic and vocal support for privatization. For a more detailed description and analysis of the political forces behind the Tax Reform Act, see Birnbaum and Murray (1987).

6 The 1981 legislation, in particular, was designed to trade a reduction in tax revenues for economic stimulation and growth. However, an economic analysis of the impact of the 1981 and 1982 tax

Acts on the treasury in the wastewater treatment area, showed no reduced revenues to the treasury (Heilman and Johnson 1992; see especially chapter 6).

7 The Deficit Reduction Act of 1984 was an interim step addressing this concern.

8 See, for example, Thelwell (1990), who concludes that while Gramm–Rudman–Hollings had some benefit in terms of making Congress more "creative" in its budget activity, the net effect of the legislation was simply to create an illusion of budget control and provide disincentives for developing meaningful deficit controls.

9 Bennett and DiLorenzo (1983, 136–137) argue that although off-budget spending has been prevalent in state and local governments for years, the federal government saw a remarkable increase in off-budget spending of 23,100 percent between 1973 and 1981, from roughly $100 million in 1973 to $23.2 billion in 1981.

10 See Palmer and Sawhill (1984, 2–10), for a discussion of the underlying Reagan philosophy, and an interpretation of the 1980 presidential election results. See also Conlan (1988, 104–106), for a further discussion of the changes in popular attitudes both before and after the 1980 election.

11 See Moe (1987), Savas (1987), and Kolderie (1986) for discussions of the various roles of the private sector in privatized arrangements. See also Johnson and Watson (1991) for an examination of the differences between the provision and the production of services in privatized arrangements.

12 Stone (1988) addresses these concepts and their attachment to both the market and the polis. See especially chapters 2–5.

13 Conlan (1988, 11–13) notes the often strong rhetoric employed by Reagan when describing the perils of big government, and the benefits of a federal system in which the federal government plays a reduced role.

14 Eads and Fix (1982) argue that one of the major underlying premises of Reagan's philosophy of government was that government regulation stifled economic productivity. Thus, the early years of Reagan's first term were marked with a great deal of activity aimed at reducing government regulation of the private sector (pp. 130–133).

15 Donald Kettl (1993, 6–10) argues that public-private partnerships in contracting for goods and services are a long-standing tradition in the United States. Furthermore, Kettl shows that many seemingly "public" functions in the areas of social policy, health policy, environmental policy, savings-and-loan bailouts, and support services, are in fact provided, produced, or delivered with a substantial private sector role. See also Zimmerman (1987) for a discussion of the effect of the 1986 Tax Reform Act on state and local revenues.

16 See Kelly (1994) for a further discussion of unfunded mandates.

17 State and local tax revolt efforts of the late 1970s and early 1980s, such as Proposition 13 in California, are indicative of the resource problems faced by states. In addition, nineteen states enacted either revenue or expenditure limitations between 1976 and 1982. See Aronson and Hilley (1986), especially chapter 12. Nathan and Doolittle (1984, 99) further suggest that the movement toward fiscal restraint in government can be traced back to the fiscal crisis of New York City in the mid-1970s.

18 Nathan and Doolittle (1984) report a mixed result in this area. Although some states in their study found Reagan's version of federalism to their liking, several states noted increased fiscal pressures brought about as a result of cuts in entitlement programs.

19 The funding role of the federal government was not to end immediately, however. The Water Quality Act of 1987 authorized $18 billion in loan capitalization grants through FY 1994 from $2.4 billion in the first year to $600 million in FY 1994. In addition, the WQA coupled loan capitalization funds with a phased withdrawal from the construction grants program. The Act authorized $2.4 billion for grants in FY 1986, 1987, and 1988, and $1.2 billion for grants in FY 1989 and 1990.

20 It must be noted, however, that the nature of the metapolicy virtually guaranteed a *lack* of coordination between state governments and the federal government. See Johnson and Heilman (1987), especially pp. 470–472, for amplification of this argument.

21 It is worth calling attention to the differences between "federalism" and "intergovernmental relations." While the terms are often used interchangeably, Wright (1988, 14) defines "intergovernmental relations" as "an important body of activities or interactions occurring

between governmental units of all types and levels within the [US] federal system," a set of activities that are highly interdependent. On the other hand, the term "federalism" tends to focus more directly on federal–state relationships, and occasionally on interstate relationships. Thus, intergovernmental relations may be thought of as a more inclusive concept, in that it takes into account relationships between governments of all types, at all levels. Although there is certainly a role for local governments in the CWSRF program, the program itself is administered by states through a grant from the national government, and the regulatory framework in place is a creation of the national government. Therefore, our attention in this work is focused on federalism, as opposed to intergovernmental relations.

22 Under the 1965 Act, states were required to follow national standards. Zimmerman (2001) argues that there were still elements of a partnership in this instance, in that states could choose to act to meet the federal standards, and to carry out the enforcement function for the national government. Only in cases in which the states refused to act would the national government preempt state law and assume responsibility for enforcement in that state.

23 As noted in Chapter 1, the Title VI grants are, for all intents and purposes, a form of block grant.

24 Cho and Wright (2001) note that Elazar (1965) also addressed this duality and called it "coercive cooperation."

25 A more cynical view of Reagan federalism suggests that Reagan's goals in federalism had nothing to do with governance *per se*, but rather was a diversionary tactic to address federal budget deficits (see Conlan 1998).

26 Wassenberg (1986) asserts that these same arguments existed prior to the development of the Clean Water Act in 1972. In short, states did not possess the technical and financial resources to assume responsibility for water quality; thus a tougher, more centralized solution to water quality was needed. Conversely, the US Advisory Commission on Intergovernmental Relations (1985) reported that state capacity had grown significantly since the 1930s. Still, the question remains: Were states in the 1980s ready and able to assume responsibility for a complex financial program?

27 A study by Heilman and Johnson (1991) conducted in the early years of the CWSRF program also reported a wide variation in state administrative capacity, specifically the degree to which CWSRF coordinators reported an adequacy of administrative, financial, and technical resources. This issue is examined in greater detail in a later chapter of this volume.

28 Although the movement toward block grants is often attributed to Reagan, the first block grant program was enacted under the Johnson administration. Block grants were also a cornerstone of Nixon's "New Federalism"; the trend continued under Reagan (Nathan and Doolittle 1987).

29 Against this backdrop, state water quality needs were also increasing, and a steady increase in the rate of inflation in this period meant that the gap between needs and resources was also increasing.

30 Morris et al. (2013) note that the groups studied in their comparative case study indicate varying degrees of success in terms of their ability to restore the watersheds in their communities. They also note (2014) that these groups can act as "force multipliers" to assist or empower local governments to address watershed restoration issues. This may be particularly important in terms of more recent requirements for states and communities to meet national Total Maximum Daily Load (TMDL) standards for pollution in waterways. The issue of TMDL requirements is covered elsewhere in this volume.

31 At the time most of this work was published, the only operational national revolving fund program was Title VI of the Water Quality Act of 1987. However, in 1996 Congress amended the Safe Drinking Water Act (SDWA), which also included a revolving fund model to provide for drinking water system infrastructure. The SDWA program took some years to develop. At this time, the term "SRF" applied to the wastewater program; as the drinking water program became more prevalent, the terms CWSRF (for clean/wastewater) and DWSRF (for drinking water) came into vogue. As a general convention, this book refers to the program from the WQA as the "CWSRF."

32 During the first six years of the program (1988–1993) states made a total of 1,902 loans, of which twenty-four loans (1.26 percent) were for nonpoint uses. Data are from EPA (2020a).

33 To be fair, the WQA also requires states to maintain the Title VI programs "in perpetuity," which clearly places an imperative on states to administer the programs in a fiscally responsible manner. In a sense, the WQA might be said to contain conflicting goals: provide resources to high-risk

communities but be sure the fund is maintained in perpetuity. States thus had to make decisions, by accident if not on purpose, about which goal to maximize. Stone (1988) offers additional thoughts on the problem of conflicting policy goals. Still, the salient point is that to turn over responsibility for the allocation of public resources to the private sector leads to very different distribution patterns.

34 I am indebted to Dr. Gerald A. Emison, Professor Emeritus at Mississippi State University, for his insights into this period in EPA history. Prior to his service in academia, Professor Emison served in the Senior Executive Service in EPA, and was personally involved in much of the work described in this section.

35 Posner and Wrightson (1996) note that the use of block grants has been closely tied to specific political agendas; they note that both Nixon and Reagan promoted the use of block grants because their use fit larger political and philosophical goals.

36 These standards generally (but not exclusively) fall into one of two categories of wastewater needs: secondary treatment (Category I) and advanced treatment (Category II).

37 Other allowable needs include stormwater needs, energy and water conservation, estuary assistance, planning and assessments, and desalination. EPA also tracks assistance to Native American tribes, but to date only twenty-one loans have been made for tribal uses.

38 It is important to note, however, that while the WQA did provide additional resources for nonpoint-source pollution, Congress elected to not impose a federal regulatory framework for nonpoint pollution, instead leaving standards and regulation to the states. Additionally, the WQA arguably brought greater attention to nonpoint pollution, but rather than being driven by more stringent policy, the renewed focus on nonpoint pollution was driven by citizen suits brought against EPA and states to enforce portions of the law already on the books.

39 For example, the 2014 amendments to the Water Resources Reform and Development Act (P.L. 113–121) required states to comply with the requirements of the so-called Buy American requirement to use American steel in their projects funded through the CWSRF.

40 The EPA working group also considered several other funding possibilities, including municipal self-funding, municipal bonds with federal credit enhancement, federal loans, and federal grants. The federal credit enhancements would involve the sale of federal bonds to secure funding for wastewater treatment. See EPA (1984, especially chapter 5).

Chapter 3

1 In this context, "policy decision" is meant to represent a tangible change in policy direction, intent, or purpose. For example, the passage of a new public law, signed by chief executive and implemented by the responsible agency is a tangible change. By contrast, a proposed bill may be defeated in the legislature or vetoed by the chief executive. In this case, the result may still represent a policy decision – let us not do anything differently that we are doing now – but does not represent a tangible change from the status quo. However, both may be rightly thought of as policy decisions: the former decision changes the status quo; the latter decision reaffirms the status quo.

2 The fight over the Hetch Hetchy dam project was one of the more controversial decisions of the period. Considered by many to be the equal of the Yosemite Valley in terms of its beauty, the Hetch Hetchy Valley and the Tuolumne River were still largely inaccessible to casual visitors and thus considered by some to be expendable. The Raker Act allowed the construction of the dam to provide a stable water supply for the San Francisco Bay area in California. The Sierra Club, which to this point was more akin to a social club than an environmental organization, was perhaps the loudest voice in opposition to the project. Led by John Muir and William Colby, the club members worked tirelessly to seek an alternative solution. However, the political forces in favor of development were too strong to overcome, and the dam flooded the valley to create the reservoir (see Cohen 1988).

3 Formally known as the "Amendments to the Federal Water Pollution Control Act of 1972," the Act is more commonly referred to as the Clean Water Act (CWA). In this chapter the terms are used interchangeably; this naming convention is explored in greater detail in Chapter 4.

4 The first municipal sewage treatment plant in the United States was constructed in 1856 (Ridgeway 1970).

5 We should note, however, that while the 1899 legislation was ineffective, later policy actions aimed specifically at the disposal of all kinds of waste from vessels have been more effective. Much of the later legislation was a result of the United States' participation in a series of international treaties, such as the Convention of the High Seas and the International Convention for the Prevention of Pollution from Ships, that were meant to regulate the dumping of waste by vessels in international waters.

6 Hugh Heclo (1978) coined the term "iron triangle" to describe the traditional power of congressional subcommittees, implementing agencies, and interest groups. The notion is that all three groups have a vested interest in their shared policy arena, and will work together to control both process and outcomes for their collective mutual benefit. Heclo suggests these "iron triangles" can dominate a policy arena, making it difficult for change to occur. Iron triangles are useful analytical tools in water quality policy, and help explain the incremental nature of change in the arena.

7 It is important to note that a separate body of policy had been developing that focused on pollution from ships. Around this same time, international conventions began to be negotiated to address ocean pollution. One of the earliest attempts was the International Convention for Prevention of Pollution in the Sea by Oil in 1954 (Schoenbaum 2019a). Perhaps the most important was the Convention on the High Seas (1958), which later became known as the United Nations Convention on the Law of the Sea (UNCLOS). Following the wreck of the oil tanker *Torrey Canyon* in March 1967 off the coast of southern England, the grounding of the tanker *Ocean Eagle* off Puerto Rico in 1968, and the oil spill off the California coast in 1968, more attention began to be paid to pollution at sea. This would eventually lead to the International Convention for the Prevention of Pollution from Ships (MARPOL), which set rules governing maritime pollution and took a more comprehensive approach to ocean pollution than had earlier attempts (Schoenbaum 2019b).

For our purposes, the policy track governing ocean pollution is parallel to, but distinct from, the policy path described in this volume. The focus of this volume, and of the Water Quality Act of 1987, is the prevention and treatment of municipal and industrial wastes originating on land. While a failure to treat this effluent can certainly have an impact on ocean environments, ocean pollution is generally addressed in law and policy through a separate policy regime.

8 As an indicator, in 1965 roughly a third of all Americans believed that water and air pollution were serious problems where they lived. By 1967 the percentage had grown to more than 50 percent, and by 1970 almost 70 percent of Americans surveyed saw these issues as significant problems. Likewise, national membership in the Sierra Club, the oldest environmental organization in the nation, nearly tripled between 1965 and 1970, reaching nearly 100,000 members in 1970 (Rinde 2017).

9 This issue is addressed in Emison and Morris (2012); see especially chapters 1 and 9.

Chapter 4

1 The term "Clean Water Act" was the short title given to the 1977 amendments to the FWPCA. The 1972 legislation provided no short title, and is thus formally the 1972 amendments to the FWPCA. However, the 1972 act was widely referred to as the "Clean Water Act" prior to 1977; indeed, the term "Clean Water Act" was used in informal discussions during the development of the legislation. In the intervening years, the convention in the literature has been to refer to the 1972 legislation as the Clean Water Act (see Adler, Landman, and Cameron 1993; see also Davis 1998, 66). To avoid confusion, this book will refer to the 1972 amendments as the "Clean Water Act."

2 Command-and-control regulatory structures rely on the legal-rational authority of government to force compliance with legal requirements. Authority is provided through legislation; federal agencies then promulgate rules and regulations governing the specific requirements of those to be regulated, and (typically) the same agency is tasked with enforcement of those same regulations. By contrast, a growing trend in the environmental arena suggests that more participative forms of governance, such as collaboration, may be more effective in the long run. While advocates of collaboration (see Morris et al. 2013; Sabatier et al. 2005) suggest that regulatory structures are limited in their effectiveness, advocates of strong command-and-control regulatory frameworks

(see Ernst 2010) believe that environmental success is dependent on the exercise of governmental coercion through the effective application of law.

3 Davies and Davies (1975) also point to the innovations in policy brought about by the passage of the Clean Air Act. Many of the same policy makers were involved in both pieces of legislation, and many of the underlying features of the CWA are the same as those in the CAA. This also serves as further evidence of the influence of the solution stream present: The solutions present at the time relied more heavily on an expanded federal role, especially in terms of regulatory enforcement.

4 The proposal also allowed for up to 10 percent of the funds to be allocated directly by the EPA administrator, although many objected to the notion of a "slush fund" for the administrator (Lieber 1975, 33).

5 The Nixon administration in 1970 had reinvigorated a little-used permit program controlled by the US Army Corps of Engineers under authority contained in the Refuse Act. This was the genesis of Nixon's move to expand the role of the Corps; Nixon hoped to integrate the Corps' function with that of EPA (Milazzo 2006, 195).

6 The language in the Act is as follows: "It is the national goal that wherever attainable, an interim goal of water quality which provides for the protection and propagation of fish, shellfish, and wildlife and provides for recreation in or on the water be achieved by July 1, 1983." This was the language that developed from the Tunney amendment, as determined by the conference committee in the fall of 1972. This is generally referred to as the "fishable and swimmable" goal in the Act (see Adler, Landman, and Cameron 1993, 8–9).

7 The Nixon administration had also recently resurrected authority under the Refuse Act of 1899 (see Eames 1970), and was using this authority to enforce water quality standards. Indeed, the evolving permitting system was grounded in this legal authority, much to the chagrin of industry. The industrial lobby was therefore wary of attempts to strengthen the power of national regulators (Lieber 1975).

8 In his veto message to Congress, Nixon heralded his record as an environmental leader, but railed against the increase in federal effort: "...the bill which has now come to my desk would provide for the commitment of a staggering, budget-wrecking $24 billion. Every extra dollar which S. 2770 contemplates spending beyond the level of my budget proposals [$6 billion] would exact a price from the consumer in the form of inflated living costs, or from the taxpayer in the form of a new Federal tax bite, or both" (CRS 1973, 138). Nixon went on to admit it was likely his veto would be overridden, but also noted that "Certain provisions of S. 2770 confer a measure of spending discretion and flexibility upon the President, and if forced to administer this legislation I mean to use those provisions to put the brakes on budget-wrecking expenditures as much as possible" (CRS 1973, 139). Nixon's veto was overridden by a vote of 52-12 in the Senate and 247-23 in the House. The votes were held on October 17–18, just weeks before the 1972 election, which likely accounts for the relatively low vote totals.

9 The CWA gave the US Coast Guard and the EPA jurisdiction over the discharge of pollution from ships in US waters. Section 301(a) of the CWA explicitly prohibits the discharge of pollutants, except in very narrowly defined circumstances. Section 312 covers discharges from commercial fishing vessels and recreational vessels, and specifies standards for marine sanitation devices (Schoenbaum 2019b). These authorities would be expanded over the course of the next two decades as the United States became a signatory to the MARPOL treaty. Subsequent legislation gave the Coast Guard authority to enforce the MARPOL requirements. Following the grounding and sinking of the *Exxon Valdez* in 1989, the Oil Pollution Act (OPA) of 1990 greatly expanded the Coast Guard's role to regulate both oil spills and oil tanker construction.

10 See National Council on Public Works Improvement (1988); see also Heilman and Johnson (1992, chapter 2) for an analysis of WTW funding patterns during the 1970s and 1980s.

11 Glicksman and Batzel (2010) suggest that the "no discharge" goal was retained because it would have been politically infeasible to remove the goal. They further suggest that the development of the National Pollutant Discharge Elimination System (NPDES) effectively negated the "no discharge" requirement as a motivating force for implementation of the Act.

12 These elements are discussed in greater detail in Chapter 2.

13 The EPA members included five assistant administrators, a regional administrator, a deputy regional administrator, EPA's general counsel, and a regional water administrator (EPA 1984, ii).

14 ASIWPCA would later change its name to the Association of Clean Water Administrators (ACWA) in August 2011.

15 The EPA received about thirty responses from a range of respondents, including state and local governments, environmental groups, engineering and financial groups, and private citizens (EPA 1984, 1-3).

16 Core needs were defined by the task force as a combination of Category I (secondary treatment), Category II (advanced treatment), Category IIIa (correction of infiltration/inflow, and Category IVB (new interceptor sewers). See EPA (1984, 3-11).

17 The five criteria were defined as follows (EPA 1984, 5-1): "Effectiveness as measured by how quickly an option will target funds to meet core treatment needs, the ability to promote capital formation for State/local self-sufficiency, and the potential flexibility in influencing long-term compliance. Efficiency as measured by the ability to allow communities meet wastewater treatment requirements at the least cost, to encourage the use of appropriate low-cost capital and O&M solutions, and to increase State flexibility in the use of funds. Equity as measured by the potential flexibility in addressing the issue of affordability across communities, and the potential fiscal impacts in Federal, State, and local budgets. Feasibility as measured by the complexity (and related cost) associated with administrative requirements to Federal, State, and local governments, and any potential negative or political of legal precedents." Again, it is clear to see that these criteria are consistent with the policy streams in place at the time.

18 One of the less obvious, but nonetheless important, elements of the task force's recommendation is that the net effect of the change to a revolving loan fund program and state self-sufficiency would be the end of EPA control over what, at the time, was EPA's largest grant program. If the program had gone as intended, EPA would have given up billions of dollars of budget authority. This is especially interesting given that the majority of the task force members were from the Office of Water, which would have been the office most directly affected by the loss of budget authority.

19 Under the Clean Water Act of 1972, the rural set-aside in Section 205(h) was intended to serve small communities, highly dispersed sections of larger municipalities, or to develop and employ "...alternatives to conventional sewage treatment..." (EPA 1984, 3 [Attachment A]). This set-aside would later be recast as a means to serve small communities, with a specific definition of community size attached.

20 One might consider the change from a categorical grants program to a short-term block grant something more than an incremental change. It is hard to argue that, at least, the change does not represent an increment larger in magnitude. However, this should be balanced against two other factors. First, the underlying regulatory framework would not change, only the mechanism through which we would fund water quality. Second, the content of the three streams (politics, problems, and solutions) had changed significantly in the previous decade. Sometimes termed the "Reagan Revolution" (see Conlan 1988), this change was real, and we would thus expect to detect larger policy changes as a result.

21 This provision is especially notable because it had the effect of exempting agricultural stormwater runoff from the regulatory regime of the NPDES. Because such runoff was not deemed a point source of pollution, it was considered nonpoint pollution and thus was not subject to regulation under the WQA.

22 The conference bill also contained a number of requirements for states to meet the needs of both small communities and financially distressed communities, as well as programs for Alaska villages and Native American reservations. These requirements (and their effects) are examined in closer detail in later chapters of this book.

23 President Reagan's veto message was eerily similar in language to the veto message issued by Richard Nixon in 1972 for the Clean Water Act. While espousing his commitment to the environment, the message also stated that "Notwithstanding my recommendations, S. 1128 would authorize $18 billion, or triple the amount I requested for that grant program, expand the allowable uses of Federal funds, and continue Federal grants for another 9 years. By 1993,

S. 1128 would increase outlays by as much as $10 billion over the projections in my 1987 budget. . ." (Office of the Federal Register 1986, 1541). Reagan also objected specifically to the $500 million authorized in the bill for nonpoint-source projects.

24 Reagan's veto message for the 1987 legislation was very similar to his statement three months earlier. Reagan began by reiterating his commitment to the environment, and again noted the cost of the legislation. He also decried federal involvement in a program ". . .that historically and properly was the responsibility of State and local governments" (Office of the Federal Register 1987, 97). However, this veto message went into more detail regarding the president's objections to the nonpoint requirements of the bill: "This new program threatens to become the ultimate whip hand for Federal regulators. For example, in participating States, if farmers have more run-off from their land than the Environmental Protection Agency decides is right, that agency will be able to intrude into decisions such as how and where the farmers must plow their fields, what fertilizers they must use, and what kind of cover crops they must plant. To take another example the Agency will be able to become a major force in local zoning decisions that will determine whether families can do such basic things as build a new home. That is too much power for anyone to have, least of all the Federal government" (Office of the Federal Register 1987, 97–98). The message was clear: While the Water Quality Act would fundamentally change the relationship between the national government and the states, Reagan still viewed the WQA as too intrusive. Reagan was a consistent advocate for federal deregulation and greater states' rights, so his position is not surprising. Indeed, it is highly reflective of the political stream in place at the time, albeit perhaps a more strident version of the message.

25 The "no" votes in both chambers were largely partisan; in the Senate, thirteen of the fourteen "no" votes were cast by Republican senators; in the House, all "no" votes were cast by Republicans. In both chambers, the "yes" votes were strongly bipartisan, an indication of widespread support for the legislation.

26 This was known as the "Buy American" requirement. The amendment did specify that this requirement could be waived if domestic supplies were unavailable, or if domestic supplies would result in a cost increase of more than 25 percent. In a survey of CWSRF state coordinators conducted in 2020 by the author, respondents were asked to name the three most important weaknesses of the CWSRF program. The "Buy American" requirements were named by nearly a quarter of respondents. More than 80 percent of all weaknesses mentioned involved federal cross-cutting requirements.

27 As this book goes to press, the president has just signed a roughly $1 trillion infrastructure package that includes some $55 billion in new spending over five years for water infrastructure. An early analysis of the legislation suggests that the water quality portion of the new funding is to be distributed and administered through the framework of the WQA.

Chapter 5

1 As noted in Chapter 1, while the Congressional Research Service does not list the CWSRF program as a block grant, the CWSRF grants share almost all characteristics of block grants. For the purposes of this research, the CWSRF program is considered a block grant.

2 Oliver Houck (2002) provides an outstanding review of the history of Section 303(d), along with an in-depth analysis of the effects of the program.

3 Section 319 is technically created by an amendment to Section 316.

4 The Intended Use Plan (IUP) identifies the uses of the funds in the CWSRF, and how those uses serve to support the goals of the CWSRF program. States prepare a new IUP annually, and the IUP must be made available for public comment and review prior to its submission to the EPA. EPA is required to have an IUP on file before the capitalization grant can be released to the state.

5 This section thus provides the legal requirement for the 20 percent state match for all federal grants made under Title VI. The 20 percent figure is mentioned at other points in the WQA, but the legal requirement is found in Section 602(b)(2).

6 It is notable that most of these requirements are identical to those proposed by the EPA working group report released in 1984. In that report, detailed in Chapter 4, the working group proposed a range of allowable uses, on the grounds that such an approach would enhance state discretion and authority (see EPA 1984).

7 This was later increased to a maximum loan term of thirty years under language included in the Water Resources Reform and Development Act of 2014 (P.L. 113–121).

8 The origins of the funding formula included in Section 205(c) are unknown (Ramseur 2016). While previous funding formulae were based on some combination of population and need, the legislative record does not include any information as to how the funding formula for the 1987 legislation was created.

9 The early years of the program were intended to provide a "bridge" between the Title II (construction grants) program and the new CWSRF program. To this end, Congress authorized $2.4 billion per year in Title II funds for FY 1986 to FY 1988 and $1.2 billion per year for FY 1989 and FY 1990 in order to give states the time necessary to pass enabling legislation at the state level to create the state programs, and to budget for the required 20 percent matching funds. This proved to be prescient; FY 1991 was the first year in which all states received federal capitalization grants. Like the CWSRF capitalization grants, Congress ultimately appropriated Title II funds well past the expiration of the authorization in the WQA; the last Title II funds were appropriated in FY 2013.

10 The adoption of private banks as an interim step in the transfer process is also reflective of Reagan's desire to involve the private sector in the business of government. This transfer requirement provides a significant revenue stream for the private banks, in that it not only increases their cash reserves, they also earn short-term interest on significantly large sums of money.

11 The letter of credit process was converted to an electronic payment process in 1979; this process connected the US Treasury, the US Federal Reserve, and federal agencies into a single network to facilitate the transfer of funds.

12 Given the formula of (federal grant)/(federal grant + state match) = federal share, we can calculate the differences. If a state only matches with the minimum amount, then we have 100/(100 + 20) = 100/120 = 0.83 (83 percent). On the other hand, if a state doubles its match to 40 percent of the capitalization grant, then the federal share drops as well: 100/(100 + 40) = 100/140 = 0.71 (71 percent). For a state that engages in leveraging, the rules allow for a state to draw up to 100 percent of the portion of the capitalization grants to be allocated to guarantee the leveraged debt.

13 As of this writing, Massachusetts, New Hampshire, and New Mexico have not assumed primacy.

14 Section 101 authorizes $135 million per year (FY 1986–FY 1990) to fund studies of the effectiveness of innovative technologies (Section 517). Section 316 (which creates the new Section 319) authorizes $70 million for FY 1988; $100 million each for FY 1989 and FY 1990; and $130 million for FY 1991, or an average of $100 million over each of the four years of the authorization.

15 In his veto statement of the Water Quality Act, Ronald Reagan made specific note of the perceived negative effect of the WQA, and specifically of the nonpoint requirements in the Act, on farmers.

Chapter 6

1 See Chapter 1 for a more detailed argument regarding the nomenclature for Title VI CWSRF grants.

2 Additional state spending might include additional dollars allocated above and beyond the state match or, more commonly, funds raised through leveraging.

3 This is similar in context to Downs' (1967) concept of the "statesman" bureaucrat, who fulfills their organizational role not on the basis of personal or organizational preference, but rather on what they believe is best for the country.

4 Informal conversations in 2019 and 2020 with former EPA regional staff indicate that the quality of engagement differs significantly across EPA regions. To this end, it seems reasonable that these differences may lead to differences in state policy choices. The relationship between states and EPA regional offices was also addressed by Heilman and Johnson (1991) and Morris (1994).

5 Although the requirement for the quadrennial needs assessment reporting appears to be still on the books, there is no evidence that reports were released in either 2016 or 2020.

6 The needs assessments capture several different kinds of need, including primary, secondary, and tertiary wastewater treatment, stormwater and combined sewer overflow needs, and several other categories. See EPA (2012) for an example of a needs assessment report.

7 A glance at the EPA data available on the agency's website (https://iaspub.epa.gov/waters10/attains_index.home) indicates the scope of the problem. First, not all states report data yearly. Second, the data reported are not in a common format; some states report results for only a few waterways, while other states report more general data. Finally, the number of waterways in a state that are surveyed varies widely. For example, Arizona reports 2.6 percent of its rivers and streams assessed in 2016, while New Hampshire reports 100 percent of rivers and streams assessed in 2012.

8 This measure is an example of a variable that was created using data from the 1980s, but has not been updated since. Although it was a significant predictor of state behavior in early cross-sectional studies, it is highly likely that relative levels of state commitment have changed. Because the variable has not been updated, it is not useful for a study that extends its observations into the second decade of a new century.

9 It is important to note that their analysis covers the time period between 1989 and 2008. However, for reasons that are unclear, the wastewater needs data are from 1988, even though needs assessments were available for 1992, 1996, 2000, 2004, and 2008.

10 As of this writing, the NIMS includes data through FY 2019. However, the data in the NIMS are subject to change for two main reasons. First, states audit their programs, and as audits change the values in the data, the changes are reported to EPA for updating. Second, because states can draw their capitalization grants over a period that extends past the fiscal year in which the funds are appropriated, the funds may not be reflected in the NIMS for up to two years after initial appropriation. A cursory inspection of NIMS data sheets over the past four fiscal years suggests that the data are typically finalized within three years. Although there are occasionally some changes after that period, the changes are generally small. We also assume that the changes are randomly distributed over all states, and with nearly 1,500 total state years of data, the effect of the changes on the models is negligible.

11 Data calculated by the author from NIMS data. States reported a total of 16,717 loans for nonpoint uses, for a total cost of $4,764,571,597, an average of $285,014 per loan. By way of comparison, states made a total of 13,791 loans for Category I or Category II needs (significant environmental need) for a total cost of nearly $63 billion, for an average loan value of $4,611,803. See Chapter 9 for additional discussion.

12 Many states house the CWSRF program in a state environmental agency. However, some states house the program in a financial agency, and still others house the program in a more traditional public health agency. This issue is discussed in further detail in Chapter 7.

13 The Clean Water Act of 1972 authorized EPA to award primacy to states. Some states achieved primacy well before the passage of the WQA in 1987; as of this writing, three states have yet to assume primacy for water quality.

14 A variable included in some earlier studies of the CWSRF program is state contribution beyond the required 20 percent match. However, an examination of the NIMS data reveals that only a handful of states have made such contributions, and no state has made additional contributions in more than five fiscal years since the inception of the CWSRF program. In total, these additional contributions amount to less than 1 percent of state match contributions. The present study excludes this variable because of the small number of nonzero observations in the dataset.

15 As of this writing, three states have yet to assume primacy.

16 An early version of the model for nonpoint resources included the NPDES variable; the variable failed to approach statistical significance in these models.

17 It should be noted that the WQA allows for the use of CWSRF funds for other uses, including estuary projects, and assistance to Indian tribes. Several other categories, including water and energy conservation, were added later as amendments to other legislation. However, an examination of the NIMS data reveals that few states have made loans for these uses. This issue is addressed again in Chapter 9.

Chapter 7

1 The principal investigators on this grant were Professors John G. Heilman and Gerald W. Johnson of Auburn University; the data are used here with their kind permission.

2 At this point in time, EPA had not yet implemented its National Information Management System for the CWSRF program. The only sources of data about program outputs were state annual reports (required under the law to be submitted by all states who had received a capitalization grant the previous year). To this end, the survey included a number of questions about dollar amounts, leveraging status, interest rate structures, and the patterns of loan distribution within the state.

3 A third survey was conducted by the author in 1996. However, that survey repeated fewer questions, and the relatively low response rate render the data from that survey less useful for these analyses.

4 Utah had passed legislation in 1983 that allowed for the operation of a similar program, and the CWSRF program operated initially under that authority. That 1983 law was not program-specific; the law was later updated.

5 Alabama represents an exception to this discussion. For the purposes of this analysis, Alabama is coded here as an "environmental agency." The enabling legislation created a five-person commission with primary legal authority for the state CWSRF program. However, the commission is, in effect, a "rubber stamp" organization; the primary program authority was actually exercised by a private sector consultant who effectively administered the program. The Alabama Department of Environmental Management (ADEM) continued to provide technical expertise, development of priority lists, and engineering consultation. This situation is discussed in greater detail in the case study chapter later in this book.

6 The discerning reader will remember that the initial federal policy instruments in the water quality arena were defined as public health policies, and until the mid-1960s, the federal water quality program was located in the Department of Health, Education, and Welfare. Many states therefore placed responsibility for their state water quality program in a health agency. With the establishment of the EPA, the federal water quality program was fully entrenched in the new environmental agency, and many states followed suit by creating state environmental agencies to mirror the federal structure. However, a number of states continued to implement their state water quality program within a state health agency.

7 State attendees at these workshops reported a clear preference by EPA officials for leveraging, and much of the financial discussion revolved around the benefits and costs of leveraging, and ways to structure and operate leveraged programs. As one attendee reported in 1990, "For EPA, leveraging was a foregone conclusion. The only real question was how to do it." EPA officials suggested the emphasis on leveraging represented an EPA conclusion that the authorization provided in the WQA was insufficient to meet water quality needs in most states without leveraging, and EPA officials had little hope that the program would be reauthorized at sufficient levels. While Congress has still not reauthorized the CWSRF program as of this writing, Congress has appropriated funds in each fiscal year since the program's inception.

8 It is worth noting that the WQA retained construction grants funding for two years as a "bridge" between the existing program and the Title VI program. States were given the option to convert their appropriated Title II money to Title VI funds if they chose. It is also notable that, despite the intent to continue the Title II program for only a short time, Congress continued to appropriate funds for the Title II program through FY 2013 (EPA 2020a).

9 EPA offered a series of training sessions around the country in 1988–1990, and invited state water pollution staff to these sessions to discuss CWSRF operation. In addition, EPA created and disseminated a range of technical materials and guidance, including a roughly 200-page management manual (EPA 1988b).

10 This is particularly true for states that operate a leveraged program. Most leveraged states charge an administrative fee to each loan or divert a portion of the loan repayment stream to offset administrative expenses.

11 The enabling legislation was both short and vague, and focused on providing legal authority for the environmental and financial agencies to implement the program. Many of the specifics were added through administrative rules and regulations after the federal legislation was passed.

12 In some respects, this point of view is not as outrageous as it might first seem. Congress had authorized Title II (construction grants) funds through FY 1990, and some states took this as an indication that the Title II program would continue. Indeed, this prediction was well-founded; Title II funds continued to be appropriated almost continuously through FY 2013.

13 The careful reader will note that managerial resources are omitted from this portion of the analysis. Of the thirty-nine states responding to this question on the survey, none of the respondents reported that managerial resources were initially insufficient. Since there is no variance in this question, the variable was omitted from the analysis. Because the survey respondents were all CWSRF state coordinators, it is likely that respondent bias accounted for the lack of variation; since the respondents *were* the managerial resources, it seems unlikely they would regard this category as insufficient.

14 When asked this same set of questions in 2020, more than 90 percent of respondents reported sufficiency for political, organizational, and managerial resources. However, nearly half of the respondents reported insufficient environmental personnel. At first glance this seems surprising, but given the tendency of states to include larger numbers of financial personnel in their CWSRF programs over time, this insufficiency is not implausible.

15 Twenty-seven states reported having a state-run and state-funded grant program. However, the resources needed to administer a grant program are different than those required for a loan-based program, so these twenty-seven states are not included in this analysis. An initial version of this analysis did include states with grant programs, but the results were not substantially different than those reported here.

16 A feature of the construction grants program, the priority list is created by the state to rank-order water quality projects based on environmental need. Although other factors (such as a readiness to proceed with the project) may be reflected in the rankings, the list is meant to be an indicator of relative environmental need of a project. Under the construction grants program, these lists were developed by the state agency responsible for the state's wastewater program.

17 This represented something of a windfall for states that were on the leading edge of program implementation. Under the WQA, if a state was not prepared to receive its capitalization grant within twelve months after the end of the authorized fiscal year, the "unclaimed" funds were redistributed to other states in the form of additional capitalization grants. The funds were distributed on the basis of the same funding formula used to determine the amount of the initial capitalization grants. In essence, this resulted in millions of "unclaimed" dollars being reallocated to states that were up and running early.

18 This role is compared to the construction grants program, which allowed virtually no private sector activity in the program's management and administration.

19 Of the forty-four valid responses received for this set of items, only six states indicated no private sector role (formal or informal) in their CWSRF program. Most states thus include at least some role for the private sector.

20 Data in this analysis were collected during the 1990 survey of CWSRF coordinators. The coordinators were asked about the involvement of private sector actors during the development stages of the SRF program (FY 1987 through FY 1990). The numbers in the table accompanying this analysis (Table 7.13) indicate the mean of the coordinators' perceptions of the level of involvement of private actors across the set of program areas listed.

21 As an example of the importance of history, during data collection for the 1991 USGS report (Heilman and Johnson 1991), the study team interviewed the state CWSRF coordinator in Texas. The interviewee was asked to explain the development of water quality administration in Texas. In the early years of federal water quality policy, the controlling agency in Texas was the Texas Railroad Commission (TRC). A commission with a remarkable degree of political power in the state, the TRC made the argument that passenger trains regularly dumped effluent from toilets on the trains directly on the tracks below. Since this pollution was a result of railroad activity, the TRC should become the lead agency for wastewater pollution.

Chapter 8

1 The interviews conducted in 1990–1991 were done so under the auspices of a grant awarded by the US Geological Survey to Dr. John G. Heilman and Dr. Gerald W. Johnson (principal investigators) at Auburn University. Also conducting interviews were Dr. Laurence J. O'Toole, this author, and Dr. Gloria (Burch) Smith. The data are also reported in Heilman and Johnson (1991) and are used with the kind permission of the principal investigators.

2 The interviews conducted in 1994 were conducted by the author as part of the dissertation process; these data are also reported in Morris (1994).

3 EPA calculated Alabama's 1988 water quality needs to be $547 million, compared with a national average of $1.32 billion (EPA 1989b, A-3).

4 Like many states, Alabama relies on federal money from the Community Development Block Grant (CDBG) program and other federal programs to fund portions of certain kinds of water quality projects.

5 The City of Auburn, AL, operated a privatized treatment plant, and was the first municipality in the nation to execute a privatization agreement for wastewater treatment services. Despite Auburn's national stature as the first municipality to privatize its treatment facilities, few other cities in Alabama followed suit. See Heilman and Johnson (1992) and Johnson and Watson (1991) for a further discussion of Auburn's privatization experiences in this area.

6 Indeed, the state legislature has never appropriated state matching funds for the CWSRF program. The match requirement is met with leveraged dollars.

7 Alabama's enabling legislation specifically limits the use of CWSRF funds to centralized wastewater treatment projects, which precludes the use of any funds for nonpoint-pollution projects.

8 Discussions with the financial advisor conducted during the 1991 CWSRF study revealed that the financial advisor's firm maintained close contact with the US Senators from Louisiana during the development of the Water Quality Act (1987). These contacts were used as a means to prepare a market opportunity in response to the potential for leveraging in the CWSRF. A pattern thus emerges in which the private sector utilized political contacts to gain a market advantage over potential competitors.

9 Alabama elects the lieutenant governor on a separate ballot, so the governor and the lieutenant governor may be political rivals. The Speaker of the House is a member of the House, and the lieutenant governor presides over the Alabama Senate. The lieutenant governor and the Speaker of the House thus represent the interests of the legislature, and the other three members represent the interests of the chief executive.

10 Alabama also ranks first in the nation in terms of freshwater biodiversity, and 10 percent of the freshwater resources in the continental US either originate in, or flow through, the state (alabamarivers.org 2020).

11 This issue is addressed more fully in the latter sections of this case study.

12 Economic analyses of the CWSRF funding model conducted by Heilman and Johnson (1991, Chapter Five) indicated that if CWSRF loans were not made above the prevailing rate of inflation, the effects of inflation would erode the long-term real value of the fund, thus eroding the future purchasing power of the fund. See also Holcombe (1992).

13 Alabama achieved primacy in 1979. Of our other cases, California assumed primacy in 1973; Georgia assumed primacy in 1974. Massachusetts is one of three states that has still not assumed primacy in the water quality program.

14 California eventually did leverage, but not until 2003. The state implemented leveraging in one other year (2016) during the study period.

15 A study conducted by Holcombe (1992) reached a similar conclusion. Even in a best-case scenario, it was likely that the available funds in the program would continue to decrease to a point that the fund would be essentially worthless in fifty years. If a state were to make loans below the prevailing rate of inflation, the value of the fund would decrease much more quickly. Holcombe's analysis assumed federal capitalization through FY 1994 (as authorized in the law); however, the federal government has appropriated funds for the CWSRF every year since.

16 California also leveraged in 2017 and 2018, raising an additional $1.05 billion in CWSRF funds. The state did not leverage in 2019.

17 This comparison is developed in a following section of this case study.

18 Initially, this duty was assumed to involve the completion of outstanding grant projects. However, Congress continued to appropriate funds under Title II (the construction grants program) through FY 2013.

19 It should be noted that Georgia does not have any sizable estuaries, and thus little opportunity to fund projects in this category.

20 The data from that survey are used with the kind permission of the authors.

21 At the time of these interviews, the authorization contained in the WQA was still in effect, and there were early discussions in Congress regarding reauthorization. The interviews conducted in 1991 were more pessimistic in terms of the long-term financial viability of the program. As it turned out the law has not yet been reauthorized, but Congress has continued to appropriate funds for capitalization grants.

Chapter 9

1 An analysis of NIMS reports posted to the EPA website between 2016 and 2019 indicates that adjustments to the NIMS data are common for the state-reported data elements. The data are generally stable after three years; for example, the 2015 data were the same for reports issued in 2018 and 2019. Not all states revise their data, and EPA revises the national report as individual states report changes. As of this writing 2016 is the most recent stable year, so the data in our analyses terminate in 2016.

2 The models for this study were estimated using several specification techniques, including OLS regression, generalized least squares, time series cross-sectional modeling, and panel-corrected regression. The alternative model specifications did not change the substantive interpretation of the findings. Estimates from the OLS regression specification are presented here as they provide the most intuitive substantive interpretation.

3 A report issued by EPA in 2018 calculated that, based on a cumulative federal capitalization of $42 billion, states had raised an additional $84 billion through leveraging. For each dollar of federal investment, $3 had been provided to communities in the form of CWSRF assistance (EPA 2018).

4 The exception is the period from 2009 to 2011, in which states committed more resources as part of the American Recovery and Reinvestment Act (ARRA), which provided additional federal funds and offered incentives for increased state investment in infrastructure projects.

5 Ten of the fifty states have never engaged in leveraging. Of states that have leveraged at least once, an additional twelve states have leveraged in ten or fewer years. Only twelve states have leveraged in twenty or more years. Data are from NIMS (EPA 2020a).

6 Refer to Figure 1.1.

7 These are defined by the EPA (2016, 13–21) as follows:

> Category I Capital costs for POTWs [Publicly-Owned Treatment Works] to meet secondary treatment standards
> Category II Capital costs for POTWs to attain a more stringent level of treatment than Cat. I
> Category III Capital costs to repair and rehabilitate sewer lines
> Category IV Capital costs to install new sewer lines, collection systems, interceptor sewers, and pumping stations
> Category V Capital costs to prevent the discharges of stormwater and untreated sewage during wet-weather events
> Category VI Measures to control the runoff of water from precipitation
> Category X Capital costs for the treatment and conveyance of water to be recycled after treatment

8 This analysis addresses national needs, rather than needs by state.

9 There have been seven needs assessments reported to Congress during the life of the Water Quality Act. For the sake of clarity, this analysis employs the initial report (1988), the most recent report (2012), and the assessment conducted at the midpoint between the two (2000).

10 The Category X needs for 2012 were reported as $6.085 billion; in 2004 the needs were reported as $6.049 billion (in 2012 dollars). These needs thus remained relatively stable between the

reporting periods. Total investment in Category X needs over the life of the CWSRF has been $916.8 million (calculated from EPA 2020a).

11 An exception to this was a separate report released by EPA in 1999 that calculated small community needs based on the 1996 Clean Water Needs Survey (EPA 1997). At the time, EPA estimated that small community needs were $13.8 billion by 2016, and that nearly 60 percent of all small community needs were concentrated in ten states (EPA 1999, 1).

12 Data for Section 319 grants are drawn from EPA (2011) and from EPA (2020b).

13 For example, the state enabling legislation in Alabama expressly limits use of CWSRF funds for traditional wastewater infrastructure.

14 This figure includes nearly $4 billion in additional funds appropriated in 2009 under the American Recovery and Reinvestment Act (ARRA).

15 Indeed, as this book goes to press, Congress just passed a large infrastructure bill that would provide $13 billion in additional water quality funding over the next five years.

16 Note that it is not possible for a loan to satisfy environmental need and nonpoint source, since they are classified as different sources of pollution.

17 This analysis does not include Davis and Lester's (1989) measure of state commitment to the environment, largely because their data were collected in the 1970s and early 1980s, and the measure has not been updated. It is likely that states have changed their stance since the original measure was developed, and it would not be a valid measure for the period covered by this study.

18 As a test of this proposition, the models were run again in stages, dropping the most recent three years of data each time. In models prior to 2007 the coefficient is positive and significant.

19 Another factor likely at play in this case is that, due to missing data for two variables of interest (privatization and initial sufficiency), Iowa was omitted from the analyses. Iowa is a moralistic state (Elazar 1966), and has made more loans for nonpoint projects than any other state. Iowa is also a state with a very strong agricultural sector. The inclusion of Iowa in the analysis would likely impact the relationship between political culture and the nonpoint-dependent variables.

20 This may be an artifact of model specification. In early analyses for this project, models using different combinations of variables generally produced a positive and significant coefficient for water quality needs. Furthermore, the positive associations between leveraging and dollars for significant environmental need suggest that states with greater water quality needs would direct more resources to significant environmental need, since the effect of leveraging is to increase the amount of money available with which to make loans. Finally, since water quality needs are positive and significant for small communities, and at least some small communities are likely to also have significant water quality needs (according to EPA (2016), more than 80 percent of all CWSRSF resources are allocated to small communities), this suggests that the lack of significance is the result of model specification error.

Chapter 10

1 Calculated from NIMS data (EPA 2020a).

2 In a 2020 survey of CWSRF coordinators, respondents were asked for their top three recommendations for changes to the CWSRF program. The most common responses were to remove two cross-cutting provisions: the "American Iron and Steel" requirement from 2014, and the Davis–Bacon wage requirements from the 1987 law.

3 Bunch (2008) reported that the lowest interest rate charged by any state on CWSRF loans was zero percent (MA and VT), with a national average of 2.4 percent. However, Heilman and Johnson (1991) reported the use of negative interest rates by several states. Unpublished survey results from surveys of CWSRF coordinators conducted by the author in 1994 and 1996 also indicated several states made loans at negative interest rates.

4 As noted in an earlier chapter, these requirements were also included in the 1972 legislation, but were largely ignored by both EPA and the states. The WQA amended these requirements to strengthen the TMDL program, but it still took more than a decade of legal action before EPA began to address the TMDL program in earnest. See Houck (2002) for a more complete discussion of this point.

5 A report issued by the US Government Accountability Office (GAO 2006) examined state efforts to address environmental outcomes in the CWSRF program. They found that while all states

complied with the financial reporting requirement of the National Information Management System (NIMS) for financial data reporting, virtually no states attempted to measure the environmental benefits of the CWSRF program. The report also noted that the WQA did not require states to monitor environmental benefits.

6 For example, the nearly $4 billion in additional funds appropriated in 2009 under the American Recovery and Reinvestment Act (ARRA) generated hundreds of new loans across the nation, which led to the creation of thousands of construction jobs.

7 Indeed, this is the example typically used by proponents of public choice approaches to governance (see, among others, V. Ostrom 1989; E. Ostrom 1990, 1991). In short, the size of the decision-making body should fit the size of the problem. Because pollution travels in ways that do not respect political boundaries, all affected jurisdictions should be both part of the solution and responsible for a single goal.

8 However, GAO has indicated the importance of local governments in the quest for national water quality standards in several reports. A 2006 report (GAO 2006, 4) acknowledged that local government bear primary responsibility for wastewater treatment, and a 2016 report found that the EPA knew little about the infrastructure needs of midsize cities facing population declines (GAO 2016). Likewise, a 2007 report suggested that little was known about the impact of EPAs stormwater program on municipalities (GAO 2007). In short, while there is awareness of the lack of information about the impacts of water quality requirements on municipalities, there has been little effort to collect and analyze data about these impacts.

9 As anecdotal evidence of this point, the 2020 survey of CWSRF state coordinators mentioned elsewhere in this volume were asked about demand for various kinds of CWSRF projects, and also asked about demand for CWSRF assistance by recipient type. The results suggested significant variation across states in both instances. Likewise, Heilman and Johnson (1991) reported that some states marketed their CWSRF programs aggressively, while other states made little or no efforts to market the program.

References

Adler, Robert Q., Jessica C. Landman, and Diane M. Cameron. 1993. *The Clean Water Act 20 Years Later*. Washington, DC: Island Press.

alabamarivers.org. 2020. "About Alabama's Watersheds." http://alabamarivers.org/about-alabamas-rivers. Accessed December 5, 2020.

American Society of Civil Engineers (ASCE). 2017. "2017 Infrastructure Report Card." www.infrastructurereportcard.org/wp-content/uploads/2017/01/Drinking-Water-Final.pdf. Accessed February 1, 2021.

Andreen, William L. 2004. "Water Quality Today: Has the Clean Water Act Been a Success?" *Alabama Law Review* 55, no. 3: 537–593.

Aronson, J. Richard, and John L. Hilley. 1986. *Financing State and Local Governments*, 4th ed. Washington, DC: Brookings Institution.

Association of Interstate Water Pollution Control Administrators (AISWPCA). 2004. *Clean Water Act Thirty-Year Retrospective: History and Documents Related to the Federal Statute*. Washington, DC: AISWPCA.

Banfield, Edward C., and James Q. Wilson. 1963. *City Politics*. Cambridge, MA: Harvard University and MIT Press.

Bennett, James T., and Thomas J. DiLorenzo. 1983. *Underground Government: The Off-Budget Public Sector*. Washington, DC: The Cato Institute.

Berry, William D., Richard C. Fording, Evan J. Ringquist, Russell L. Hanson, and Carl E. Klarner. 2010. "Measuring Citizen and Government Ideology in the U.S. States: A Re-appraisal." *State Politics & Policy Quarterly* 10, no. 2: 117–135.

Berry, William D., Evan J. Ringquist, Richard C. Fording, and Russell L. Hanson. 1998. "Measuring Citizen and Government Ideology in the American States, 1960–93." *American Journal of Political Science* 42, no. 1: 327–348.

Birnbaum, Jeffrey H., and Alan S. Murray. 1987. *Showdown at Gucci Gulch: Lawmakers, Lobbyists, and the Unlikely Triumph of Tax Reform*. New York: Vintage Books.

Bowen, Daniel, and Zachary Greene. 2014. "Should We Measure Professionalism with an Index? A Note of Theory and Practice in State Legislative Professionalism Research." *State Politics & Policy Quarterly* 14, no. 3: 277–296.

Bowling, Cynthia J., Jonathan M. Fisk, and John C. Morris. 2020. "Seeking Patterns in Chaos: Transactional Federalism in the Trump Administration's Response to the COVID-19 Pandemic." *American Review of Public Administration* 50, nos. 6–7: 512–518.

Bowling, Cynthia J., and J. Mitchell Pickerill. 2013. "Fragmented Federalism: The State of American Federalism 2012–13." *Publius: The Journal of Federalism* 43, no. 3: 315–346.

Bowling, Cynthia J., and Deil S. Wright. 1998. "Change and Continuity in State Administration: Administrative Leadership across Four Decades." *Public Administration Review* 58, no. 5: 429–444.

Bowman, Ann, and Richard C. Kearney. 1988. "Dimensions of State Government Capacity." *Western Political Quarterly* 41, no. 2: 341–362.

Bowman, Ann, and Richard C. Kearney. 2011. "Second-Order Devolution: Data and Doubt." *Publius: The Journal of Federalism* 41, no. 4: 563–585.

Breaux, David A., Gerald A. Emison, John C. Morris, and Rick Travis. 2010. "State Commitment to Environmental Quality in the South: A Regional Analysis." In *Speaking Green with a Southern Accent: Environmental Management and Innovation in the South*, edited by Gerald A. Emison and John C. Morris, 19–33. Lanham, MD: Lexington Books.

Bressers, Hans Th. A. 1992. "The 'Cleaning Up Period' in Dutch Water Quality Management: Policy Instruments Assessed." In *International Comparative Policy Research: Preparing a Four Country Study on Water Quality Management*, edited by Hans Bressers and Laurence J. O'Toole Jr., 155–174. Enschede, The Netherlands: Center for Clean Technology and Environmental Policy, University of Twente.

Bressers, Hans Th. A., and Pieter Jan Klok. 1988. "Fundamentals for a Theory of Policy Instruments." *International Journal of Social Economics* 15, no. 3/4: 22–41.

Bullock, Charles S. III, and Mark J. Rozell, eds. 2018. *The New Politics of the Old South: An Introduction to Southern Politics*, 6th ed. Lanham, MD: Rowman & Littlefield.

Bulman-Pozen, Jessica. 2014. "Partisan Federalism." *Yale Law Journal* 127, no. 4: 1077–1145.

Bunch, Beverly S. 2008. "Clean Water State Revolving Fund Program: Analysis of Variations in State Practices." *International Journal of Public Administration* 31, no. 2: 117–136.

Cannon, Lou. 2003. *Governor Reagan: His Rise to Power*. New York: PublicAffairs.

Caraley, Demetrios, and Yvette R. Schlussel. 1986. "Congress and Reagan's New Federalism." *Publius: The Journal of Federalism* 16, no. 1: 49–79.

Carson, Rachel. 1962. *Silent Spring*. Boston: Houghton Mifflin.

Chamberlain, Adam. 2013. "The (Dis)Connection between Political Culture and External Efficacy." *American Politics Research* 41, no. 5: 761–782.

Cho, Chung-Lae, and Deil S. Wright. 2001. "Managing Carrots and Sticks: Changes in State Administrators' Perceptions of Cooperative and Coercive Federalism during the 1990s." *Publius: The Journal of Federalism* 31, no. 2: 57–80.

Clark, Benjamin, and Andrew B. Whitford. 2010. "Does More Federal Environmental Funding Increase or Decrease States' Efforts?" *Journal of Policy Analysis and Management* 30, no. 1: 136–152.

Clynch, Edward J. 1972. "A Critique of Ira Sharkansky's 'The Utility of Elazar's Political Culture'" *Polity* 5, no. 1: 139–141.

Cohen, Michael P. 1988. *The History of the Sierra Club, 1892–1970*. San Francisco: Sierra Club Books.

Cohen, Michael, James G. March, and Johann P. Olsen. 1972. "A Garbage Can Model of Organizational Choice." *Administrative Science Quarterly* 17, no. 1: 1–25.

Coleman, Arica L. 2016, July 31. "When the NRA Supported Gun Control." Time.com. https://time.com/4431356/nra-gun-control-history/. Accessed March 17, 2020.

Coleman Battista, James, and Jesse T. Richman. 2011. "Party Pressures in the US State Legislatures." *Legislative Studies Quarterly* 36, no. 3: 397–422.

Colman, William G. 1989. *State and Local Government and Public-Private Partnerships: A Policy-Issues Handbook*. New York: Greenwood Press.

Congressional Research Service. 1973. *A Legislative History of the Water Pollution Control Act Amendments of 1972 – Vol. I*. Washington, DC: US Government Printing Press (Serial No. 93-1).

Conlan, Timothy. 1986. "Federalism and Competing Values in the Reagan Administration." *Publius: The Journal of Federalism* 16, no. 1: 29–47.

Conlan, Timothy. 1988. *New Federalism: Intergovernmental Reform from Nixon to Reagan*. Washington, DC: Brookings Institution.

Conlan, Timothy. 1998. *From New Federalism to Devolution: Twenty-Five Years of Intergovernmental Reform*. Washington, DC: Brookings Institution.

Conlan, Timothy. 2006. "From Cooperative to Opportunistic Federalism: Reflections on the Half-Century Anniversary of the Commission on Intergovernmental Relations." *Public Administration Review* 66, no. 5: 663–676.

Cooper, Christopher A., and H. Gibbs Knotts. 2017. *The Resilience of Southern Identity*. Chapel Hill, NC: UNC Press.

Copeland, Claudia. 2014. *Wastewater Treatment: Overview and Background*. Washington, DC: Congressional Research Service, 98–323.

Corwin, Edward S. 1950. "The Passing of Dual Federalism." *Virginia Law Review* 36, no. 1: 1–24.

Cox, Gary W., Thad Kousser, and Mathew D. McCubbins. 2010. "Party Power or Preferences? Quasi-Experimental Evidence from American State Legislatures." *The Journal of Politics* 72, no. 3: 799–811.

Crotty, Patricia M. 1987. "The New Federalism Game: Primacy Implementation of Environmental Policy." *Publius: The Journal of Federalism* 17, no. 2: 53–67.

Crotty, Patricia M. 1988. "Assessing the Role of Federal Administrative Agencies: An Exploratory Analysis." *Public Administration Review* 48, no. 2: 642–648.

Daley, Dorothy M., and James C. Garand. 2005. "Horizontal Diffusion, Vertical Diffusion, and Internal Pressure in State Environmental Policymaking, 1989–1998." *American Politics Research* 35, no. 5: 615–644.

Daley, Dorothy M., Megan Mullin, and Meghan Rubado. 2013. "State Agency Discretion in a Delegated Federal Program: Evidence from Drinking Water Investment." *Publius: The Journal of Federalism* 44, no. 4: 564–586.

Davies, J. Clarence, and Barbara S. Davies. 1975. *The Politics of Pollution*, 2nd ed. Indianapolis, IN: Bobbs-Merrill.

Davis, Charles E., and James P. Lester. 1989. "Federalism and Environmental Policy." In *Environmental Politics and Policy: Theories and Evidence*, edited by James P. Lester, 57–84. Durham, NC: Duke University Press.

Davis, Charles E., and Sandra K. Davis. 1999. "State Enforcement of the Federal Hazardous Waste Program." *Polity* 31, no. 3: 451–468.

Davis, David Howard. 1998. *American Environmental Politics*. Chicago: Nelson Hall.

Derthick, Martha. 1996. "Whither Federalism?" Urban Institute, "The Future of the Public Sector," no. 2. http://webarchive.urban.org/UploadedPDF/derthick.pdf. Accessed January 9, 2021.

Dilger, Robert J. 1986. "Grantsmanship, Formulamanship, and Other Allocational Principles: Wastewater Treatment Grants." In *American Intergovernmental Relations Today: Perspectives and Controversies*, edited by Robert J. Dilger, 147–167. Englewood Cliffs, NJ: Prentice Hall.

Downs, Anthony. 1967. *Inside Bureaucracy*. Boston: Little Brown.

Dror, Yehezkel. 1968. *Public Policymaking Reexamined*. Scranton, PA: Chandler.

Eads, George C., and Michael Fix. 1982. "Regulatory Policy." In *The Reagan Experiment: An Examination of Economic and Social Policies under the Reagan Administration*, edited by John L. Palmer and Isabel V. Sawhill, 129–153. Washington, DC: The Urban Institute Press.

Eames, Diane D. 1970. "The Refuse Act of 1899: Its Scope and Role in Control of Water Pollution." *California Law Review* 58: 1444–1473.

Elazar, Daniel J. 1962. *The American Partnership: Intergovernmental Co-operation in the Nineteenth-Century United States*. Chicago: University of Chicago Press.

Elazar, Daniel J. 1965. "The Shaping of Intergovernmental Relations in the Twentieth Century." *Annals of the American Academy of Political and Social Science* 359: 10–22.

Elazar, Daniel J. 1966. *American Federalism: A View from the States*. New York: Harper and Row.

Emison, Gerald A., and John C. Morris. 2012. *True Green: Executive Effectiveness in the U.S. Environmental Protection Agency*. Lanham, MD: Lexington Books.

Environmental Financial Advisory Board (EFAB). 2008. *Report on the Relative Benefits of the Direct Loan and Leveraged Loan Approaches for Structuring State Revolving Loan Funds*. Washington, CD: EFAB. https://nepis.epa.gov/Exe/ZyNET.exe/ P100AA7K.txt?ZyActionD=ZyDocument&Client=EPA&Index=2006%20Thru%20 2010&Docs=&Query=&Time=&EndTime=&SearchMethod=1&TocRestrict=n&Toc= &TocEntry=&QField=&QFieldYear=&QFieldMonth=&QFieldDay=&UseQField=& IntQFieldOp=0&ExtQFieldOp=0&XmlQuery=&File=D%3A%5CZYFILES%5 CINDEX%20DATA%5C06THRU10%5CTXT%5C00000025%5CP100AA7K.txt& User=ANONYMOUS&Password=anonymous&SortMethod=h%7C-&Maximum Documents=1&FuzzyDegree=0&ImageQuality=r75g8/r75g8/x150y150g16/i425& Display=hpfr&DefSeekPage=x&SearchBack=ZyActionL&Back=ZyActionS&Back Desc=Results%20page&MaximumPages=1&ZyEntry=3&slide. Accessed January 12, 2021.

Erikson, Robert S., Gerald C. Wright Jr., and John P. McIver. 1989. "Political Parties, Public Opinion, and State Policy in the United States." *American Political Science Review* 83, no. 3: 729–750.

Ernst, Howard R. 2010. *Fight for the Bay: Why a Dark Green Environmental Awakening Is Needed to Save the Chesapeake Bay*. Lanham, MD: Rowman & Littlefield.

Flatt, Victor B. 1997. "Dirty River Runs through It (The Failure of Enforcement in the Clean Water Act)." *Boston College Environmental Affairs Law Review* 25, no. 1: 1–45.

Formisano, Ronald P. 2001. "The Concept of Political Culture." *Journal of Interdisciplinary History* 31, no. 3: 393–426.

Fowler, Luke, and Chris Birdsall. 2020. "Does the Primacy System Work? State versus Federal Implementation of the Clean Water Act." *Publius: The Journal of Federalism* 51, no. 1: 131–160. doi: 10.1093/publius/pjaaa011.

Freeman, A. Myrick III. 1990. "Water Pollution Policy." In *Public Policies for Environmental Protection*, edited by Paul Portney, 97–149. Washington, DC: Resources for the Future.

Gerlak, Andrea K. 2006. "Federalism and US Water Policy: Lessons for the Twenty-First Century." *Publius: The Journal of Federalism* 36, no. 2: 231–257.

Glicksman, Robert L., and Matthew R. Batzel. 2010. "Science, Politics, and the Arc of the Clean Water Act: The Role of Assumptions in the Adoption of a Pollution Control Landmark." *Washington University Journal of Law & Policy* 32: 99–138.

Grumm, John. 1971. "The Effects of Legislative Structure on Legislative Performance." In *State and Urban Politics*, edited by Richard I. Hofferbert and Ira Sharkansky, 298–322. Boston: Little Brown.

Heclo, Hugh. 1978. "Issue Networks and the Executive Establishment." In *The New American Political System*, edited by Anthony King. Washington, DC: American Enterprise Institute, pp. 87–124.

Heilman, John G., and Gerald W. Johnson. 1991. *State Revolving Loan Funds: Analysis of Institutional Arrangements and Distributive Consequences*. Auburn University, AL: Final Report Submitted to the U.S. Geological Survey, Department of the Interior.

Heilman, John G., and Gerald W. Johnson. 1992. *The Politics and Economics of Privatization: The Case of Wastewater Treatment*. Tuscaloosa, AL: University of Alabama Press.

Heilman, John G., Gerald W. Johnson, John C. Morris, and Laurence J. O'Toole Jr. 1994. "Water Policy in the United States." *Environmental Politics* 3 (special issue): 80–109.

Heritage Foundation. 1987. "Building Grass-Roots Support for Privatization." Presentation by Rep. Curt Weldon to the Heritage Foundation, May 19. Published in *Heritage Lectures*, no. 112. Washington, DC: Heritage Foundation.

Holcombe, Randall G. 1992. "Revolving Fund Finance: The Case of Wastewater Treatment." *Public Budgeting & Finance* 12, no. 3: 50–65.

Hoornbeek, John A. 2005. "The Promises and Pitfalls of Devolution: Water Pollution Policies in the American States." *Publius: The Journal of Federalism* 35, no. 1: 87–114.

Hoornbeek, John A. 2011. *Water Pollution Policies and the American States: Runaway Bureaucracies or Congressional Control?* Albany, NY: SUNY Press.

Houck, Oliver A. 2002. *The Clean Water Act TMDL Program: Law Policy, and Implementation*. Washington, DC: Environmental Law Institute.

Hugo, Victor. [1862] 1931. *Les Miserables*. Translated by Charles E. Wilbour. New York: The Modern Library.

Hunter, Susan, and Richard W. Waterman. 1996. *Enforcing the Law: The Case of the Clean Water Acts*. Armonk, NY: M.E. Sharpe.

Jaroscak, Joseph V., Julie M. Lawhorn, and Robert J. Dilger. 2020. "Block Grants: Perspectives and Controversies." Washington, DC: Congressional Research Service, R40486.

Johnson, Craig L. 1995. "Managing Financial Resources to Meet Environmental Infrastructure Needs: The Case of State Revolving Funds." *Public Productivity & Management Review* 18, no. 3: 263–275.

Johnson, Gerald W., and John G. Heilman. 1987. "Metapolicy Transition and Policy Implementation: New Federalism and Privatization." *Public Administration Review* 47, no. 6: 468–78.

Johnson, Gerald W., and Douglas J. Watson. 1991. "Privatization: Provision or Production of Services?" *State and Local Government Review* 23, no. 2: 82–89.

Jones, Bryan D., and Frank R. Baumgartner. 2005. *The Politics of Attention: How Government Prioritizes Problems*. Chicago: University of Chicago Press.

Kelly, Janet. 1994. "Unfunded Mandates: The View from the States." *Public Administration Review* 54, no. 4: 405–408.

Kettl, Donald F. 1988. *Government by Proxy: (Mis?)Managing Federal Programs*. Washington, DC: Congressional Quarterly Press.

Kettl, Donald F. 1993. *Sharing Power: Public Governance and Private Markets*. Washington, DC: Brookings Institution.

Kettl, Donald F. 2020. *The Divided States of America: Why Federalism Doesn't Work*. Princeton, NJ: Princeton University Press.

Kincaid, John. 1990. "From Cooperative to Coercive Federalism." *The Annals of the American Academy of Political and Social Science* 509, no. 1: 139–152.

Kingdon, John W. 1984. *Agendas, Alternatives, and Public Policies*. New York: Harper Collins.

Klyza, Christopher M., and David J. Sousa. 2013. *American Environmental Policy: Beyond Gridlock*. Cambridge, MA: MIT Press.

Kneeland, Douglas E. August 4, 1980. "Reagan Campaigns at Mississippi Fair; Nominee Tells Crowd of 10,000 He Is Backing States' Rights." *The New York Times*. p. A11. Accessed March 17, 2020.

Knight, Chris. 2013. "Court Rulings Highlight EPA's Mixed Record on 'Cooperative Federalism'." *Inside EPA's Clean Air Report* 24, no. 19 (September 12): 36–37.

Kolderie, Ted. 1986. "The Two Different Concepts of Privatization." *Public Administration Review* 46, no. 4: 285–291.

Kraft, Michael E., and Norman J. Vig. 1990. "Environmental Policy from the Seventies to the Nineties." In *Environmental Policy in the 1990s*, edited by Norman E. Vig and Michael E. Kraft, 1–30. Washington, DC: Congressional Quarterly Press.

Lasswell, Harold. 1936. *Politics: Who Gets What, When, and How*. Chicago: University of Chicago Press.

Lees, John David, and Michael Turner. 1988. *Reagan's First Four Years: A New Beginning?* Manchester: Manchester University Press.

Lester, James P. 1986. "New Federalism and Environmental Policy." *Publius: The Journal of Federalism* 16, no. 1: 149–165.

Lester, James P. ed. 1989. *Environmental Politics and Policy: Theories and Evidence*. Durham, NC: Duke University Press.

Lieber, Harvey. 1975. *Federalism and Clean Waters: The 1972 Water Pollution Control Act*. Lexington, MA: Lexington Books.

Lindblom, Charles E. 1959. "The Science of Muddling Through." *Public Administration Review* 19, no. 2: 79–88.

List, John A., and Shelby Gerking. 2000. "Regulatory Federalism and Environmental Protection in the United States." *Journal of Regional Science* 40, no. 3: 453–471.

Lowry, William R. 1992. *The Dimensions of Federalism: State Governments and Pollution Control Policies*. Durham, NC: Duke University Press.

Lurie, Irene. 1997. "Temporary Assistance for Needy families: A Green Light for States." *Publius: The Journal of Federalism* 27, no. 2: 73–87.

McKee, Seth. 2018. *The Dynamics of Southern Politics: Causes and Consequences*. Thousand Oaks, CA: Sage.

Maxwell, Angie, and Todd Shields. 2019. *The Long Southern Strategy: How Chasing White Voters in the South Changed American Politics*. New York: Oxford University Press.

Milazzo, Paul Charles. 2006. *Unlikely Environmentalists: Congress and Clean Water, 1945–1972*. Lawrence: University of Kansas Press.

Millimet, Daniel L. 2003. "Assessing the Empirical Impact of Environmental Federalism." *Journal of Regional Science* 43, no. 4: 711–733.

Moe, Ronald C. 1987. "Exploring the Limits of Privatization." *Public Administration Review* 47, no. 6: 453–460.

Montjoy, Robert S., and Laurence J. O'Toole Jr. 1984. "Interorganizational Policy Implementation: A Theoretical Perspective." *Public Administration Review* 44, no. 6: 491–503.

Morris, John C. 1994. Privatization and Environmental Policy: An Examination of the Distributive Consequences of Private Sector Activity in State Revolving Funds. Doctoral dissertation, Auburn University, AL.

Morris, John C. 1996. "Institutional Arrangements in an Age of New Federalism: Public and Private Management in the State Revolving Fund Program." *Public Works Management & Policy* 1, no. 2: 145–157.

Morris, John C. 1997. "The Distributional Impacts of Privatization in National Water Quality Policy." *Journal of Politics* 59, no. 1: 56–72.

Morris, John C. 1999a. "Was Reagan Federalism Good for Business? Privatization and State Administrative Capacity in the Clean Water State Revolving Fund Program." *Southeastern Political Review* 27, no. 2: 323–339.

Morris, John C. 1999b. "State Implementation of National Water Quality Policy: Policy Theory, Policy Streams and (Un)Intended Consequences in the State Revolving Fund Program." *Public Works Management & Policy* 3, no. 2: 317–330.

Morris, John C. 2007. "The Artist as Environmentalist: Ansel Adams, Policy Entrepreneurship, and the Growth of Environmentalism." *Public Voices* 9, no. 2: 9–24.

Morris, John C., William A. Gibson, William M. Leavitt, and Shana C. Jones. 2013. *The Case for Grassroots Collaboration: Social Capital and Ecosystem Restoration at the Local Level*. Lanham, MD: Lexington Books.

Morris, John C., William A. Gibson, William M. Leavitt, and Shana C. Jones. 2014. "Collaborative Federalism and the Emerging Role of Local Nonprofits in Water Quality Implementation." *Publius: The Journal of Federalism* 44, no. 3: 499–518.

Morris, John C., and John G. Heilman. 2002. "Of Time and Policy Streams: The Impact of Privatization on Longitudinal Evaluation." *Journal of Global Awareness* 3, no. 1: 12–26.

Morris, John C., Ryan D. Williamson, and A. Stanley Meiburg. 2021. "Contours of a National Infrastructure Policy for the New Millennium." *Public Works Management & Policy* 26, no. 3: 200–209.

Morris, John C., Ryan D. Williamson, Lien Nguyen, and Jan C. Hume. 2021. "Promise and Performance: The Water Quality Act at Thirty." Paper presented at the Annual Meetings of the Southern Political Science Association, January 2021.

Mullin, Megan, and Dorothy M. Daley. 2017. "Multilevel Instruments for Infrastructure Investment: Evaluating State Revolving Funds for Water." *Policy Studies Journal* 46, no. 3: 629–650.

Murray, Robert K., and Tim H. Blessing. 1993. *Greatness in the White House*. State College, PA: Penn State Press.

Nathan, Richard P., and Fred C. Doolittle. 1984. "The Untold Story of Reagan's 'New Federalism'." *Public Interest* 77: 96–105.

Nathan, Richard P., and Fred C. Doolittle. 1987. *Reagan and the States*. Princeton, NJ: Princeton University Press.

National Council of State Legislatures (NCSL). 2020. "State Partisan Composition." www .ncsl.org/research/about-state-legislatures/partisan-composition.aspx. Accessed December 13, 2020.

National Council on Public Works Improvement (NCPWI). 1988. *Fragile Foundations: A Report on America's Public Works*. Washington, DC: NCPWI.

Nice, David C. 1987. *Federalism: The Politics of Intergovernmental Relations*. New York: St. Martin's Press.

Northrup, Herbert R. 1984. "The Rise and Demise of PATCO." *Industrial and Labor Relations Review* 37 no. 2: 167–184.

Office of the Federal Register. November 10, 1986. *Weekly Compilation of Presidential Documents, Vol. 22, No. 45*. Washington, DC: US Government Printing Office.

Office of the Federal Register. February 2, 1987. *Weekly Compilation of Presidential Documents, Vol. 23, No. 4*. Washington, DC: US Government Printing Office.

Office of Municipal Pollution Control (OMPC). 1988. *Letter of Credit: How Is It Used in EPA's State Revolving Fund Program?* Washington, DC: Author. EPA #430/09-88-009.

Ostrom, Elinor. 1990. *Governing the Commons: The Evolution of Institutions for Collective Action*. New York: Cambridge University Press.

Ostrom, Elinor. 1991. "A Method of Institutional Analysis and an Application to Multiorganizational Arrangements." In *The Public Sector: Challenge for Coordination and Learning*, edited by Franz-Xaver Kaufmann, 501–523. New York: Walter de Gruyter.

Ostrom, Vincent. 1989. *The Intellectual Crisis in American Public Administration*. Tuscaloosa: University of Alabama Press.

O'Sullivan, Elizabethann, and Gary R. Rassel. 1995. *Research Methods for Public Administrators*, 2nd ed. White Plains, NY: Longman.

O'Toole, Laurence J., Jr., and Robert K. Christensen. 2013. "American Intergovernmental Relations: An Overview." In *American Intergovernmental Relations: Foundations, Perspectives, and Issues*, 5th ed., edited by Laurence J. O'Toole, Jr. and Robert K. Christensen, 1–32. Los Angeles, CA: Sage.

Owings, Stephanie, and Rainald Borck. 2000. "Legislative Professionalism and Government Spending: Do Citizen Legislators Really Spend Less?" *Public Finance Review* 28, no. 3: 210–225.

Palmer, John L., and Isabel V. Sawhill. 1984. "Overview." In *The Reagan Record: An Assessment of America's Changing Domestic Priorities*, edited by John L. Palmer and Isabel V. Sawhill, 1–30. Washington, DC: The Urban Institute.

Pedersen, William F., Jr. 1988. "Turning the Tide on Water Quality." *Ecology Law Quarterly* 15, no. 1: 69–102.

Peters, Gerhard, and John T. Woolley. 2020. "Election of 1980." Santa Barbara, California: The American Presidency Project. www.presidency.ucsb.edu/documents/presiden tial-documents-archive-guidebook/list-vice-presidents-who-served-acting-president. Accessed March 17, 2020.

Peterson, George E. 1984. "Federalism and the States: An Experiment in Decentralization." In *The Reagan Record: An Assessment of America's Changing Domestic Priorities*, edited by John L. Palmer and Isabel V. Sawhill, 217–259. Cambridge, MA: Ballinger Publishing.

Poe, Gregory L. 1995. *The Evolution of Federal Water Pollution Control Policies*. Ithaca, NY: Department of Agricultural, Resource, and Managerial Economics.

Posner, Paul L. 1998. *The Politics of Unfunded Mandates: Whither Federalism?* Washington, DC: Georgetown University Press.

Posner, Paul L., and Margaret T. Wrightson. 1996. "Block Grants: A Perennial, but Unstable, Tool of Government." *Publius: The Journal of Federalism* 26, no. 3: 87–108.

President's Commission on Privatization. 1988. *Privatization: Toward More Effective Government*. Washington, DC: President's Commission on Privatization.

Pressman, Jeffrey L., and Aaron Wildavsky. 1973. *Implementation: How Great Expectations in Washington Are Dashed in Oakland; Or, Why It's Amazing That Federal Programs Work at All, This Being a Saga of the Economic Development Administration as Told by Two Sympathetic Observers Who Seek to Build Morals on the Foundation of Ruined Hopes*. Berkeley, CA: University of California Press.

Ramseur, Jonathan. 2016. *Allocation of Wastewater Treatment Assistance: Formula and Other Changes*. Washington, DC: Congressional Research Service. CRS Report RL31073 (Version 24, Updated).

Rawat, Pragati, and John C. Morris. 2019. "Kingdon's 'Streams' Model at Thirty: Still Relevant in the 21st Century?" *Politics & Policy* 44, no. 4: 608–638.

Reagan, Ronald. 1965. *Where's the Rest of Me?* New York: Duell, Sloan, and Pearce.

Ridgeway, James. 1970. *The Politics of Ecology*. New York: E.P. Dutton.

Rinde, Meir. 2017. "Richard Nixon and the Rise of Environmentalism." www
.sciencehistory.org/distillations/magazine/richard-nixon-and-the-rise-of-american-
environmentalism. Accessed December 10, 2020.

Ringquist, Evan. 1993. *Environmental Protection at the State Level: Politics and Progress
in Controlling Pollution*. Armonk, NY: M.E. Sharpe.

Rittel, Horst W. J., and Melvin M. Webber. 1973. "Dilemmas in a General Theory of
Planning." *Policy Sciences* 4: 155–169.

Rogers, Peter. 1993. *America's Water: Federal Roles and Responsibilities*. Cambridge,
MA: MIT Press.

Rosenbaum, Walter A. 2002. *Environmental Politics and Policy*, 5th ed. Washington, DC:
CQ Press.

Sabatier, Paul A., Will Focht, Mark Lubell, Zev Trachtenberg, Arnold Vedlitz, and Marty
Matlock, eds. 2005. *Swimming Upstream: Collaborative Approaches to Watershed
Management*. Cambridge, MA: MIT Press.

Savage, Robert L. 1981. "Looking for Political Subcultures: A Critique of the Rummage-
Sale Approach." *Western Political Quarterly* 34, no. 2: 331–336.

Savas, E. S. 1987. *Privatization: The Key to Better Government*. Chatham, NJ: Chatham
House.

Schlesinger, Joseph A. 1955. "A Two-Dimensional Scheme for Classifying the States
According to Degree of Inter-Party Competition." *American Political Science
Review* 49, no. 4: 1120–1128.

Schlitz, Timothy D., and R. Lee Rainey. 1978. "The Geographic Distribution of Elazar's
Political Subcultures among the Mass Population: A Research Note." *Western
Political Quarterly* 31, no. 3: 410–415.

Schoenbaum, Thomas J. 2019a. "The International Framework for Marine Pollution
Control: Admiralty and Maritime Law." *Westlaw*, §18:1 (6th ed.), 3.

Schoenbaum, Thomas J. 2019b. "Marine Pollution from Shipping Activities: An
Overview; Admiralty and Maritime Law." *Westlaw*, §18:2 (6th ed.), 1.

Shannon, John, and James E. Kee. 1989. "The Rise of Competitive Federalism." *Public
Budgeting & Finance* 9, no. 4: 5–20.

Shapiro, Michael. 1990. "Toxic Substances Policy." In *Public Policies for Environmental
Protection*, edited by Paul Portney, 195–242. Washington, DC: Resources for the
Future.

Sharp, Elaine B. 1990. *Urban Politics and Administration: From Service Delivery to
Economic Development*. New York: Longman Press.

Shimshack, Jay P. 2014. "The Economics of Environmental Monitoring and Enforcement."
Annual Review of Resource Economics 6: 339–360.

Simon, Herbert. 1948. *Administrative Behavior: A Study of Decision-Making Processes in
Administrative Organization*, 3rd ed. New York: The Free Press.

Squire, Peverill. 2007. "Measuring State Legislative Professionalism: The Squire Index
Revisited." *State Politics & Policy Quarterly* 7, no. 2: 211–227.

Squire, Peverill. 2017. "A Squire Index Update." *State Politics & Policy Quarterly* 17, no.
4: 361–377.

Stephens, G. Ross. 1974. "State Centralization and the Erosion of Local Autonomy."
Journal of Politics 36, no. 1: 44–76.

Stern, Duke E., and Edward M. Mazze. 1974. "Federal Water Pollution Control Act
Amendments of 1972." *American Business Law Journal* 12, no. 1: 81–86.

Stone, Deborah A. 1988. *Policy Paradox and Political Reason*. Glenview, IL: Scott,
Foresman.

Swartz, Thomas R., and John E. Peck. 1990. "The Changing Face of Fiscal Federalism." *Challenge* 33, no. 6: 41–46.

Tannenwald, Robert. 1999. "Fiscal Disparity Among the States Revisited." *New England Economic Review*, July/August: 3–24.

Thelwell, Raphael. 1990. "Gramm-Rudman-Hollings Four Years Later: A Dangerous Illusion." *Public Administration Review* 50, no. 2: 190–198.

Travis, Rick, and Elizabeth D. Morris. 2003. "Privatization in State Agencies: A Focus on Clean Water." *Public Works Management & Policy* 7, no. 4: 243–255.

Travis, Rick, Elizabeth D. Morris, and John C. Morris. 2004. "State Implementation of Federal Environmental Policy: Explaining Leveraging in the Clean Water State Revolving Fund." *Policy Studies Journal* 32, no. 3: 461–480.

Truman, David B. 1951. *The Governmental Process: Political Interests and Public Opinion*. New York: Knopf.

U.S. Advisory Commission on Intergovernmental Relations. 1985. *The Question of State Government Capability*. Washington, DC: Government Printing Office (January 1985).

U.S. Bureau of Labor Statistics. 2020. "Consumer Price Index (CPI) Databases." www.bls.gov/cpi/data.htm. Accessed September 9, 2020.

U.S. Census Bureau. 2020. "American Community Survey Data, 2013–2017." www.census.gov/programs-surveys/acs/data.html. Accessed August 2, 2020.

U.S. Environmental Protection Agency (EPA). 1983. *Final Report of the State/Federal Roles Task Force*. Washington, DC: Office of Policy and Resource Management, September 16, 1983.

U.S. Environmental Protection Agency (EPA). 1984. *Study of the Future Federal Role in Municipal Wastewater Treatment: Report to the Administrator*. Washington, DC: USEPA.

U.S. Environmental Protection Agency (EPA). 1988a. *SRF: Initial Guidance for State Revolving Funds*. Washington, DC: EPA Office of Water.

U.S. Environmental Protection Agency (EPA). 1988b. *State Water Pollution Control Revolving Fund: Management Manual*. Washington, DC: EPA Office of Water.

U.S. Environmental Protection Agency (EPA). 1989a. *SRF Update*. Washington, DC: USEPA Office of Water.

U.S. Environmental Protection Agency (EPA). 1989b. *1988 Needs Survey Report to Congress: Assessment of Needed Publicly Owned Wastewater Treatment Facilities in the United States*. Washington, DC: USEPA (EPA 430/09-89-001).

U.S. Environmental Protection Agency (EPA). 1993. *1992 Needs Survey Report to Congress: Assessment of Needs for Publicly Owned Wastewater Treatment Facilities, Correction of Combined Sewer Overflows, and Management of Storm Water and Nonpoint Source Pollution in the United States*. Washington, DC: USEPA (September).

U.S. Environmental Protection Agency (EPA). 1997a. "25th Anniversary of the Clean Water Act, Minneapolis, MN. Text of speech delivered by EPA Administrator Carol Browner on October 17, 1997." https://archive.epa.gov/epapages/newsroom_archive/speeches/872d86a1679743df8525701a0052e3a5.html. Accessed November 26, 2020.

U.S. Environmental Protection Agency (EPA). 1997b. *EPA 1996 Clean Water Needs Survey: Report to Congress*. Washington, DC: USEPA (832-R-97-003).

U.S. Environmental Protection Agency (EPA). 1999. *EPA 1996 Clean Water Needs Survey: Small Community Wastewater Needs*. Washington, DC: USEPA (832-F-99-058).

U.S. Environmental Protection Agency (EPA). 2003. *Clean Watersheds Needs Survey 2000: Report to Congress*. Washington, DC: USEPA. www.epa.gov/cwns/cwns-2000-report-congress. Accessed April 11, 2020.

U.S. Environmental Protection Agency (EPA). 2008. *Clean Watersheds Needs Survey 2004: Report to Congress*. Washington, DC: USEPA. www.epa.gov/cwns. Accessed April 11, 2020.

U.S. Environmental Protection Agency (EPA). 2010. *Clean Watersheds Needs Survey 2008: Report to Congress*. Washington, DC: USEPA. EPA-832-R-10-002.

U.S. Environmental Protection Agency (EPA). 2011. *A National Evaluation of the Clean Water Act Section 319 Program*. Washington, DC: Office of Wetlands, Oceans, and Watersheds. www.epa.gov/nps/319-grant-reports-and-project-summaries. Accessed July 7, 2020.

U.S. Environmental Protection Agency (EPA). 2016. *Clean Watersheds Needs Survey 2012: Report to Congress*. Washington, DC: USEPA. EPA 830-R-15005 (January 2016).

U.S. Environmental Protection Agency (EPA). 2018. *2017 Annual Report: Clean Water State Revolving Fund Programs*. Washington, DC: Author. EPA 830-R-17007.

U.S. Environmental Protection Agency (EPA). 2019. *2019 Annual Report: Building the Project Pipeline: Clean Water State Revolving Fund*. Washington, DC: Office of Wastewater Management. EPA 83-R-20001. www.epa.gov/sites/production/files/2020-10/documents/2019_cwsrf_annual_report_9-10.pdf. Accessed February 1, 2021.

U.S. Environmental Protection Agency (EPA). 2020a. "Clean Water State Revolving Fund (CWSRF) National Information Management System Reports." www.epa.gov/cwsrf/clean-water-state-revolving-fund-cwsrf-national-information-management-system-reports. Accessed March 1, 2020.

U.S. Environmental Protection Agency (EPA). 2020b. "319 Grant Program for States and Territories." www.epa.gov/nps/319-grant-program-states-and-territories. Accessed October 14, 2020.

U.S. Environmental Protection Agency (EPA). 2021a. "EPA Announces $6 Billion in New Funding for Water Infrastructure Projects." www.epa.gov/newsreleases/epa-announces-6-billion-new-funding-water-infrastructure-projects. Accessed February 1, 2021.

U.S. Environmental Protection Agency (EPA). 2021b. "Learn about the Clean Water State Revolving Fund (CWSRF)." www.epa.gov/cwsrf/learn-about-clean-water-state-revolving-fund-cwsrf. Accessed January 28, 2021.

U.S. Environmental Protection Agency (EPA). 2021c. 303(d) "Listed Impaired Waters." www.epa.gov/ceam/303d-listed-impaired-waters. Accessed January 17, 2021.

U.S. General Accounting Office (now Government Accountability Office). 1995. *Block Grants: Characteristics, Experience, and Lessons Learned*. Washington, DC: USGAO. HEHS-95-74.

U.S. Government Accountability Office (GAO). 2006. *Clean Water: How States Allocated Revolving Loan Funds and Measure Their Benefits*. Washington, DC: USGAO. GAO-06-579.

U.S. Government Accountability Office (GAO). 2007. *Clean Water: further Implementation and Better Cost Data Needed to Determine Impact of EPA's Storm Water Program on Communities*. Washington, DC: USGAO (GAO-07-479).

U.S. Government Accountability Office (GAO). 2016. *Water Infrastructure: Information on Selected Midsize and Large Cities with Declining Populations*. Washington, DC: USGAO (GAO-16-785).

Vawter, Wallace. 1955. "Water Supply and Use." Reprinted in HSCRH, *Water Pollution Control Act*, 84th Congress, 1st Session, 20 July 1955, pp. 113–129.

Veasey, R. Lawson. 1988. "Devolutionary Federalism and Elazar's Typology: The Arkansas Response to Reagan's New Federalism." *Publius: The Journal of Federalism* 18, no. 1: 61–77.

Vig, Norman J., and Michael E. Kraft, eds. 2019. *Environmental Policy: New Directions for the Twenty-First Century*. Thousand Oaks, CA: Sage.

Walker, David B. 1991. "American Federalism from Johnson to Bush." *Publius: The Journal of Federalism* 21, no. 1: 105–119.

Walker, David B. 2000. *The Rebirth of Federalism*, 2nd ed. New York: Chatham House.

Wassenberg, Pinky. 1986. "State Responses to Reductions in Federal Funds: Section 106 of the Federal Water Pollution Control Act Amendments of 1972." In *Administering the New Federalism*, edited by Lewis G. Bender and James A. Stever, 226–247. Boulder, CO: Westview Press.

Watson, Douglas J., and Thomas Vocino. 1990. "Changing Intergovernmental Fiscal Relationships: Impact of the 1986 Tax Reform Act on State and Local Governments." *Public Administration Review* 50, no. 4: 427–434.

Williamson, Ryan D., John C. Morris, and Jonathan M. Fisk. 2021. "Institutional Variation, Professionalization, and State Implementation Choices: An Examination of Investment on Water Quality across the 50 States." *American Review of Public Administration* 51, no. 6: 436–448.

Woods, Neal D. 2005. "Primacy Implementation of Environmental Policy in the U.S. States." *Publius: The Journal of Federalism* 36, no. 2: 259–276.

Woodward, Colin. 2011. *American Nations: A History of the Eleven Rival Regional Cultures of North America*. New York: Viking.

Wright, Deil S. 1988. *Understanding Intergovernmental Relations*, 3rd ed. Pacific Grove, CA: Brooks Cole Publishing.

Yager, Edward M. 2006. *Ronald Reagan's Journey: Democrat to Republican*. Lanham, MD: Rowman & Littlefield.

Zimmerman, Dennis. 1987. "Tax Reform and the State and Local Sector." *Public Administration Review* 47, no. 6: 510–514.

Zimmerman, Joseph F. 1991. "Federal Preemption under Reagan's New Federalism." *Publius: The Journal of Federalism* 21, no. 1: 7–28.

Zimmerman, Joseph F. 2001. "National-State Relations: Cooperative Federalism in the Twentieth Century." *Publius: The Journal of Federalism* 31, no. 2: 15–30.

Zuckert, Catherine H. 1983. "Reagan and that Unnamed Frenchman (De Tocqueville): On the Rationale for the New (Old) Federalism." *The Review of Politics* 45, no. 3: 421–442.

Index

Printed in the United States
by Baker & Taylor Publisher Services